5999720

16 —

158

DATE DUE

BC
135 The Logical Enterprise
L586

STOCKTON STATE COLLEGE LIBRARY
Pomona, N. J. 08240

DEMCO

THE LOGICAL ENTERPRISE

CONTRIBUTORS

Alan Ross Anderson
John Bacon
Nuel D. Belnap, Jr.
Haskell B. Curry
David Harrah
John T. Kearns
Carl R. Kordig
Stephan Körner
Ruth Barcan Marcus
R. M. Martin
John Myhill
Richard J. Orgass
W. V. O. Quine
Richmond H. Thomason
Bas C. van Fraassen
Kathleen Johnson Wu

THE LOGICAL ENTERPRISE

ALAN ROSS ANDERSON,
RUTH BARCAN MARCUS, AND
R. M. MARTIN, EDITORS

New Haven and London
Yale University Press
1975

Published with assistance from the Louis Stern
Memorial Fund.
Copyright © 1975 by Yale University.
All rights reserved. This book may not be
reproduced, in whole or in part, in any form
(except by reviewers for the public press),
without written permission from the publishers.
Library of Congress catalog card number: 74-20084
International standard book number: 0-300-01790-1
Designed by Sally Sullivan
and set in Press Roman type.
Printed in the United States of America by
The Murray Printing Co., Forge Village, Mass.
Published in Great Britain, Europe, and Africa by
Yale University Press, Ltd., London.
Distributed in Latin America by Kaiman & Polon,
Inc., New York City; in India by UBS Publishers'
Distributors Pvt., Ltd., Delhi; in Japan by John
Weatherhill, Inc., Tokyo.

STOCKTON STATE COLLEGE LIBRARY
POMONA, NEW JERSEY 08240

Dedicated to
FREDERIC B. FITCH

Contents

Preface

The papers in this volume are written in honor of Frederic B. Fitch, by friends, present or former colleagues, and former students. They all attest to the high esteem in which he is held by logicians and philosophers alike. They are dedicated to him with the kind of affection and gratitude so gentle, courteous, and helpful a person has inspired in all those who have been associated at all closely with him.

Fitch's work exhibits an unusual breadth. It includes, on the more technical side, bold and original studies in the philosophy and methodology of science, and on combinatory logic, modal logic, the problem of consistency, mathematical intuitionism, and inductive probability. The important series of papers concerned with "basic logic" and its consistency has overflowed into extensions and applications not only to denumerabilist mathematics but to the theory of sequential circuits and other engineering matters as well. Much of the motivation of this technical work, however, has been philosophical, even metaphysical. Here Fitch has been both analytic and speculative, concerned with problems in the philosophy of language, value theory, jurisprudence, and philosophical theology. Throughout his work there has been a kind of Whiteheadian *basso ostinato* connecting it with some of the scientific, metaphysical, and theological aspects of process philosophy.

A philosopher so deeply devoted to philosophy and so highly competent in logic as Fitch will inevitably regard the latter as providing the proper approach to the former. To the general reader it is perhaps this facet of Fitch's work that is the most accessible. Whatever we do in philosophy, we should do it logistically, we should make it a part of the logical enterprise, he seems to say. Even more generally, he wrote some years ago, "I believe that symbolic logic will eventually make possible new advances in value theory, epistemology, psychology, and philosophy which will be comparable with the advances that traditional mathematics, especially after the advent of calculus, has made possible in the natural sciences." The various papers of this book help to show that in the whirligig of time this belief is being justified.

R.M.M.

PART I METAPHYSICS AND LANGUAGE

1: W. V. O. QUINE On the Individuation of Attributes

For a while I propose to treat tolerantly of attributes, or properties. Usually I have taken a harsher line. Classes, down the years, I have grudgingly admitted; attributes not. I have felt that if I must come to terms with Platonism, the least I can do is keep it extensional. For this brief space, however, it will be convenient to keep the question of the existence of attributes in abeyance, or even to talk as if they existed.

Attributes are classes with a difference. That is, corresponding to one and the same class there may be several *different* attributes. The point is that no two classes have exactly the same members, but two different attributes may be attributes of exactly the same things. Classes are identical when their members are identical; such is the principle of individuation of classes. On the other hand attributes, I have often complained, have no clear principle of individuation.

Faulty individuation has nothing to do with vagueness of boundaries. We are accustomed to tolerating vagueness of boundaries. We have little choice in the matter. The boundaries of a desk are vague when we get down to the fine structure, because the clustering of the molecules grades off; the allegiance of any particular peripheral molecule is indeterminate, as between the desk and the atmosphere. However, this vagueness of boundaries detracts none from the sharpness of our *individuation* of desks and other physical objects. What the vagueness of boundaries amounts to is this: there are many almost identical physical objects, almost coextensive with one another, and differing only in the inclusion or exclusion of various peripheral molecules. Any one of these almost coextensive objects could serve as the desk, and no one the wiser; such is the vagueness of the desk. Nevertheless they *all* have their impeccable principle of individuation; physical objects are identical if and only if coextensive. Where coextensiveness is not quite fully *verifiable,* neither is identity, but the identity is still well *defined,* though the desk is not. Specification is one thing, individuation another. Physical objects are well individuated, whatever else they are not. We know what it takes to distinguish them, even where we cannot detect it.

It then follows that the classes of physical objects are well individuated too, since their identity consists simply in the identity of the members. But what, on the other hand, is the principle of individuation of attributes? Thus, grant me that any creature with a heart has kidneys and vice versa. The class of hearted creatures and the class of kidneyed creatures are identical. Still, we are not prepared to identify the attribute of heartedness with that of kidneyedness. Coextensiveness of attributes is not enough. What more, then, is wanted?

I forget who it was–let me call him Zedsky–that offered an interestingly exasperating answer to this question. The necessary and sufficient condition for identity of attributes is quite as clear, he contended, as the notions of class and member–notions with which I evidently have no quarrel. For we have only to look to Russell's general definition of identity, which explains identity for attribututes and anything else. In general, objects x and y are identical if and only if they are members of just the same classes. So attributes, in particular, are identical if and only if they are members of exactly the same classes of attributes. This of course is an exasperating defense; but it is interesting to consider why.

We might object to Zedsky's answer in the following way. He says that attributes are identical when the classes that they belong to are identical. But when are such classes identical? They are classes of attributes. They are identical when their members are identical; that was the principle of individuation of classes. It was a good principle of individuation of classes of physical objects, since we had a good prior standard of identity of physical objects. But it is useless for classes of attributes; we cannot appeal to identity of the members, failing a prior standard of individuation of attributes, these being the members. Zedsky is evidently caught in a circle, individuating attributes in terms of identity of classes whose individuation depends on that of attributes.

Zedsky might answer in turn by protesting that the Russellian definition which he is applying to attributes does not really mention identity of classes. When expressed in words it seems to do so: attributes are identical when the classes that they belong to are identical. But it is simpler in symbols, and says nothing of identity of classes. What is required in order that attributes A and B be identical is simply that $(z)\ (A \epsilon z .\equiv. B \epsilon z)$. This formula contains only the single variable 'z' for classes of attributes, and no mention of identity.

What this shows is that we must look a little deeper. The real reason why the formula does not clarify the individuation of attributes is not that it mentions identity of classes of attributes, but that it mentions classes of attributes at all. We have an acceptable notion of class, or physical object, or attribute, or any other sort of object, only insofar as we have an acceptable principle of individuation for that sort of object. There is no entity without identity. But the individuation of classes of attributes depends, we saw, on the individuation of attributes. This, then, is why we are not satisfied with an account of the individuation of attributes which, like Zedsky's, depends on the notion of classes of attributes at all.

These thoughts on attributes remind us that classes themselves are satisfactorily individuated only in a relative sense. They are as satisfactorily individuated as their members. Classes of physical objects are well individuated; so also, therefore, are classes of classes of physical objects; and so on up. Classes of attributes, on the other hand, are as badly off as attributes. The notion of a class of things makes no better sense than the notion of those things.

We may do well, with this relativism in mind, to cast an eye on the credentials of set theory. It turns out that the usual systems of set theory still stand up to the demands of individuation very well, if we assume that the ground elements or individuals of the system are physical objects or other well-individuated things rather than ill-individuated things such as attributes. For, we saw how classes of well-individuated things are well individuated; therefore so are classes of such classes; and so on up.

This takes care of the usual systems of set theory, which exclude *ungrounded* classes. A class is ungrounded if it has some member which has some member which . . . and so on downward *ad infinitum,* never reaching bottom. The system of my "New Foundations" does have ungrounded classes, and so does the system of my *Mathematical Logic;* and it could be argued that for such classes there is no satisfactory individuation. They are identical if their members are identical, and these are identical if *their* members are identical, and there is no stopping. This, then, is a point in favor of the systems that bar ungrounded classes.

Our reflections on individuation have brought out a curious comparison of three grades of stringency. We can never quite specify the desk, from among various nearly coincident physical objects, because of vagueness at the edges. Yet its individuation, we saw, was quite in order. And we saw further that though with Zedsky we can define identity for attributes by applying Russell's general definition, still the individuation of attributes is not in order. These examples suggest that specification makes the most stringent demands, individuation is less stringent, and mere definition of identity is less stringent still.

I think I have adequately explained why Zedsky's account of the identity of attributes does not solve the problem of their individuation. But it may be worthwhile still, for the sake of a shift of perspective, to answer Zedsky again along a somewhat different line. For this we must consider sentences. How do we ever actually specify any particular attribute *or* any particular class? Or, for that matter, any particular proposition? Ordinarily, basically, we do so by citing an appropriate sentence. For specifying an attribute or a class it will be an open sentence, like '*x* has a heart,' giving the attribute of heartedness and the class of the hearted. For specifying a proposition it will be a closed sentence. Attributes, classes, and propositions are what we may call, in a word, *epiphrastic.* I am not saying that these entities are somehow ontologically dependent on sentences, nor even that sentences can be formulated to fit all attributes or classes or propositions. I am saying how one ordinarily proceeds when one does succeed, if at all, in specifying any one attribute or class or proposition.

This being the case, the question of individuation of attributes becomes in practice a question of how to tell whether two open sentences express the same attribute. How should the two sentences be related? We may object to Zedsky that while his citation of Russell defines identity of attributes, it does now say how two sentences should be related in order to express an identical attribute. However, this objection needs tightening; he can rise to the new demand. By an easy adaptation of Russell's definition he can tell us how two sentences should be related. Thus let us represent the two sentences as '*Fx*' and '*Gx*', open sentences as they are. For '*Fx*' and '*Gx*' to express the same attribute, then, Zedsky can tell us, all that is required is that the attribute of being an x such that Fx, and the attribute of being an x such that Gx, belong to the same classes of attributes. In this rejoinder, Zedsky would simply be repeating the frustrating old Russell definition, but stretching it so as to squeeze the sentences 'Fx' and 'Gx' in, and thus meet our demand for a relation between sentences. Sentences 'Fx' and 'Gx' express the same attribute, so Zedsky would be telling us, just in case those sentences are so related that $(z)\,(x[Fx] \in z .\equiv. x[Gx] \in z)$.

What can we say to this? We have a pat answer, which we can carry over from the earlier argument. This is unsatisfactory in individuating attributes because, again, it assumes prior intelligibility of the notion of a class z of attributes. Indeed this is doubly unsatisfactory, for it assumes the notion of attribute also in the abstraction notation '$x[\ \ ]$', 'the attribute of being an x such that.'

So we now see that when we ask for a relation of sentences that will individuate attributes, we must require that the relation be expressed without mention of attributes. The notion of attribute is intelligible only insofar as we already know its principle of individuation.

Observe, in contrast, how well the corresponding requirement is met in the individuation of classes. I began by saying that classes are identical when their members are identical; but what we now want is a satisfactory formulation of a relation between two open sentences 'Fx' and 'Gx' which holds if and only if 'Fx' and 'Gx' determine the same class. The desired formulation is of course immediate: it is simply '$(x)\,(Fx \equiv Gx)$'. It does not talk of classes; it does not use class abstraction nor epsilon, and it does not presuppose classes as values of variables. It is as pure as the driven snow. Classes, whatever their foibles, are the very model of individuation on this approach.

Can we do anything similar for attributes? Some say we can. Open sentences 'Fx' and 'Gx' express the same attribute, they say, if and only if they entail each other; if and only if the biconditional formed from them is analytic. Or, to put the matter into modal logic, the requirement is that (x) nec$(Fx \equiv Gx)$. Thus formulated, the requirement for identity of attributes is the same as the requirement for the identity of classes, but with the necessity operator inserted. For my part, however, I find this sort of account unsatisfactory because of my doubts over the notion of necessity, or analyticity—to say nothing of modal logic.

Propositions are of course on a par with attributes. Two closed sentences are said to express the same proposition when they entail each other; when their biconditional is analytic, or holds necessarily. And I find the account similarly unsatisfactory.

I have contrasted classes with attributes in point of individuation. But I must stress that this is their only contrast. I must stress this fact because I have lately come to appreciate, to my surprise, that an old tendency still persists among philosophers to regard classes and attributes as radically unlike. It persists as an unarticulated feeling. I have been unable to elicit clear theses on the point, but I think I have sensed the association of ideas.

There seems to be an unarticulated feeling that a class is best given by enumeration. For large classes this is impossible, granted; but still it is felt that you know a class only insofar as you know its members. But may we not say equally that you know an attribute only insofar as you know what things have it? Well, the two cases are felt to differ. One cause of this feeling is perhaps an unconscious misreading of the principle that a class is "determined" by its members: as if to say that to find the class you must find its members. All it really means, for classes to be determined by their members, is that classes are the same that have the same members.

There seems, moreover, to be an unarticulated feeling that a class of widely dispersed objects is itself widely extended in the world, unwieldy, and hard to envisage. One cause of this feeling is perhaps an unconscious misreading of 'extension' or 'extensional'. The feeling is encouraged, and also evinced, by the use of such words as 'aggregate' to explain the notion of class, as if classes were concrete heaps or swarms. Indeed, the word 'class' itself, as if to say a fleet of ships, originated in this attitude. Likewise the word 'set'.

In contrast to all this, there is no reluctance to recognize that an attribute is normally specified by presenting an open sentence or predicate. The readiness to recognize this for attributes, despite not recognizing it for classes, is due perhaps to an unconscious tendency to confuse attributes with locutions, predicates. Indeed 'predicate' has notoriously been used in both senses, as was 'propositional function'. Attributes are thus associated with sentential conditions, while classes are associated with rosters or with heaps and swarms.

I am driven thus to psychological speculation as my only means of combatting the false contrast of class and attribute, because this misconception does not stand forth as an explicit thesis to refute. It is a serious misconception still, for it leads some philosophers to prefer attributes to classes, as somehow clearer or more natural, even when they do not propose to distinguish between coextensive attributes. Let us recognize rather that if we are always to count coextensive attributes as identical, attributes *are* classes. Let us recognize that classes, like attributes, are abstract and immaterial, and that classes, like attributes, are epiphrastic. You specify a class not by its members but by its membership condition, its open sentence. Any proper reason to prefer attributes to classes must hinge only on distinctions between coextensive attributes.

Such distinctions are indeed present and called for in modal logic. Also, as we saw, the acceptance of modal logic carries with it the acceptance of a way of individuating attributes. Anyone who rejects this way of individuating attributes also rejects modal logic, as I do. Now it may be supposed that the only use of attributes is in modal logic. If this supposition is true, there is no individuation problem for attributes; either we accept the individuation given by modal logic or we reject modal logic and so have no need of attributes. However, this supposition is uncertain; there may be a call for attributes independent of modal logic. Perhaps they are wanted in a theory of causality. In this event the problem of the individuation of attributes remains with us.

In an essay of fifteen years ago called "Speaking of Objects," I raised the question whether we might just acquiesce in the faulty individuation of attributes and propositions by treating them as twilight entities, only real enough to be talked of in a few limited contexts, excluding the identity context. The plan is rather reminiscent of Frege on propositional functions. For his propositional functions may well be seen as attributes, and he accorded them only a shadowy existence: they were *ungesättigt.* It is a question of moderating the maxim "No entity without identity" to allow half-entity without identity. I raised the question in passing and remarked only that such a course would require some refashioning of logic. I propose now to explore the matter a bit further.

If we continue to speak of attributes as members of classes, then we cannot dodge the question of identity of attributes. For, given this much, we can coherently ask whether attributes A and B belong to all the same classes, and to ask this is to ask whether they are identical, even failing any *satisfactory* principle of individuation. If, therefore, we are to waive individuation for attributes, we must disallow class membership on the part of attributes.

In set theory, then, we find a possible candidate for the office of attribute, namely, *ultimate classes.* The phrase is mine, but the notion goes back to von Neumann, 1925. Some historians claim to find it already in Julius König, 1905, and even in Cantor, 1899; but this hinges on a question of interpretation. Anyway, an ultimate class is a class that does not belong to any classes. In *Mathematical Logic* I called them non-elements. Often in the literature they are called *proper* classes. This perverse terminology has its etiology: they are classes *proper* as opposed to sets. But it *is* perverse; 'ultimate' is more suggestive.

The admitting of ultimate classes is one of the various ways of avoiding the antinomies of set theory, such as Russell's paradox. In the face of ultimate classes we can no longer ask, with Russell, about the class of all those classes that are not members of themselves. Since ultimate classes are members of none, the nearest we can come to Russell's paradoxical class is the class of all those *sets,* or non-ultimate classes, that are not members of themselves. This is an ultimate class, and not a member of itself, and there is no contradiction.

Such was the purpose of ultimate classes. Russell avoided the antinomies by means rather of his theory of types, of course, and did not have the notion of an ultimate class. Still this notion induces, retroactively, some nostalgic reverberations also in Russell's writings. Away back in *Principles of Mathematics* Russell was exercised over the class as many and the class as one. The ultimate class, had he seen it, he might have seen as a class that existed only as many and not as one. Or, again, recall Russell's introduction to the second edition of Whitehead and Russell's *Principia Mathematica.* There, influenced by Wittgenstein's *Tractatus,* Russell entertained the maxim that "a propositional function occurs only through its values." This echoes Frege on the *ungesättigt.* It is not clear how to make a coherent general doctrine of this maxim, without giving up general set theory. But the maxim does nicely fit the notion of an ultimate class. Being capable only of having members and not of being members, ultimate classes may be likened to propositional functions that occur only through their values.

Ultimate classes had as their purpose the avoidance of the antinomies, but perhaps we can now put them to another purpose: to serve as attributes. Ultimate classes are incapable of being members, and so are not captured by identity as Russell defines it; and identity conditions were all that stood between class and attribute. Reviving Russell's phrases, we can say that an attribute is a class as many, while a set is a class as one. An attribute is a propositional function that occurs only through its values.

Russell's definition of identity, however, now needs reconsideration. Membership on the part of ultimate classes has not been degrammaticized; it is merely denied. Hence Russell's definition of identity *can* be applied to ultimate classes, and it gives an absurd result: ultimate classes are all members of exactly the same classes, namely none, and are hence all identical—despite differing in their members. But here we are merely faced with the known and obvious fact that that definition of identity does not work for theories that admit objects belonging to no classes.

There *is* a way of constructing a suitable definition of identity within *any* theory, even a theory in which there is no talk of class and member at all. It is the method of exhaustion of the lexicon of predicates. A trivial example will remind you of how it runs. Suppose we have just two primitive one-place predicates 'F' and 'G', and one two-place predicate 'H', and no constant singular terms or functors; just quantifiers and truth functions. Then we can define '$x = y$' as:

$$Fx \equiv Fy \cdot Gx \equiv Gy \cdot (z)\,(Hxz \equiv Hyz \cdot Hzx \equiv Hzy),$$

thus providing substitutivity in atomic contexts. The full logic of identity can be derived. The method obviously extends to any finite lexicon of primitive predicates, and it defines, every time, genuine identity or an indistinguishable facsimile: indistinguishable within the terms of the theory concerned.

We have just now run a curious course. In order to shield attributes from the identity question, we tried to shield them from membership in classes, such being Russell's definition of identity. But in so doing, and thus likening attributes to ultimate classes, we have disqualified Russell's definition of identity and thus nullified our motivation.

In conclusion let us briefly consider, then, how attributes must fare under our generalized approach to identity by exhaustion of predicates. Let us vaguely assume a rich, all-purpose lexicon of predicates, no mere trio of predicates as in the last example. Certain of these predicates will be desirable in application to attributes. One such is the two-place predicate 'has'; we want to be able to speak of an object as having an attribute. Many other predicates will be useless in application to attributes; thus it would be false, at best, to affirm, and useless, at best, to deny, that an attribute is pink or divisible by four. Ryle branded such predications as category mistakes; he declared them meaningless and so did Russell in his theory of types. So did Carnap.

Over the years I have represented a minority of philosophers who preferred the opposite line: we can simplify grammar and logic by minimizing the number of our grammatical categories and maximizing their size. Instead of agreeing with Carnap that it is meaningless to say "This stone is thinking about Vienna," and with Russell that it us meaningless to say "Quadruplicity drinks procrastination," we can accommodate these sentences as meaningful and trivially false. Stones simply never think, as it happens, and quadruplicity never drinks.

For our present inquiry into attributes, however, we are bound to take rather the line of Ryle, Russell, and Carnap. We must separate the serious predications from the silly ones. We must think in terms of a many-sorted logic with many sorts of variables, some of which are grammatically attachable to some predicates and some to others. Then we can look to that partial lexicon of just those predicates that are attachable to attribute variables and to attribute names; and we can proceed to define identity of attributes by exhaustion of just those predicates. If we keep those predicates to just the ones that may seriously and usefully be affirmed of attributes, and denied of attributes, then the identity relation defined by exhaustion of those predicates should be just right for attributes.

Stated less cumbersomely, what we are seeking as an identity relation of attributes is a relation which assures interchangeability, *salva veritate,* in all contexts that are worthwhile for attribute variables. Let us try, then, to name a few such contexts. One conspicuous one we have named is the context 'has'. On the other hand a context that is still unwelcome, presumably, is the membership context: we do not want to recognize the question of membership of attributes in classes. For if membership of attributes in classes were accepted, we could again apply Russell's shortcut definition of identity, and so again we would have identity of attributes without any instructive principle of individuation. For that matter, even the 'has' context is presumably welcome only when the mention of attributes is restricted to the right-hand side; for we do not want to recognize attributes of attributes. If we recognized them, we could say that attributes are identical when they have all the same attributes, and so again we would have identity of attributes without any instructive principle of individuation. For the same reason, of course, a primitive predicate of identity would be an unwelcome context for attribute variables; and so would contexts where attributes are counted. Thus we should be prepared, at least until solving our identity problem, to abstain from saying that Napoleon had all the attributes of a great general save one—though remaining free to say that Napoleon had all the attributes of a great general.

This last example, of course, mentions attributes only in the same old 'has' context. *Are* there other desirable contexts, or are all desirable contexts of attribute variables parasitic on 'has' and reducible to it by paraphrase?

In the latter event, our definition of identity of attributes by exhaustion of lexicon is the work of a moment. The relevant lexicon is meager indeed, having 'has' as its only member; and thus attributes come out identical if exactly the same things *have* them. In this event, attributes are extensional; we might as well read 'has' as membership, and call attribute classes; but they are classes as many, not as one, for we are declaring it ungrammatical to represent them as members of further classes. They occur only through their values. They are ultimate classes, except that now we do not deny their membership in other classes; we degrammaticize it. Some set-theorists, Bernays among them, have already taken this line with ultimate classes too.

If on the other hand there are desirable contexts of attribute variables that do not reduce to 'has', then let us have a list of them in the form of a list of appropriate primitive predicates of attributes. Given the list, we know how to define identity of attributes in terms of it. It should be possible to produce the list, and thus to individuate attributes, if attributes really serve any good purpose not served by classes.

2: STEPHAN KÖRNER

On Some Relations between Logic and Metaphysics

Among the problems which have never ceased to engage the attention of Frederic Fitch are the relation between logic and metaphysics, the nature of metaphysical arguments, and the interpretation of modal logic. This essay, written in his honor, discusses the first and second of these interconnected problems in some detail and touches upon the third. It falls into two parts, of which one is mainly devoted to an analysis of various kinds of necessary connections, the other to the use made of them in metaphysical arguments. More particularly, in Part I distinctions are drawn between logical and metaphysical principles (§1), between primary and secondary logical principles (§2), and between formal and material necessity and related notions in both, the sphere of factual and the sphere of practical thinking (§3). Part II examines and illustrates metaphysical arguments based on formal and material necessity in the factual and practical spheres (§4), and metaphysical arguments based on plausibility and, hence, partly on formal and material possibility (§5). It concludes by drawing attention to various more or less obvious fallacies which occur in metaphysical reasoning (§6).

I

1. On Logical and Metaphysical Principles

A person's metaphysical principles include propositions expressing the manner in which, having differentiated his experience or world into particulars and general characteristics, he (i) classifies the particulars into *summa genera* or maximal kinds which—apart from possible common borderline cases—are exclusive and exhaustive of all particulars; (ii) associates with each maximal kind (a) certain constitutive principles, i.e. propositions to the effect that being a member of this kind logically implies having certain (constitutive) attributes as well as (b) certain individuating principles, i.e., propositions to the effect that being a distinct member of this kind logically implies possessing one or more (individuating) attributes; and (iii) distinguishes between independent particulars or substances, which exist independently of other particulars, and dependent particulars, which exist only as features of independent ones. The formulation of these metaphysical principles, some of which may be transcategorial in the sense of being associated with two or more maximal kinds, presupposes (iv) the adoption of certain logical principles. It is convenient to say that a person's principles of categorization, his constitutive and individuating principles, his principles for distinguishing substances from other particulars, and his logical principles, demarcate his "categorial framework." It will, moreover, be convenient to distinguish the logical principles (of propositional logic, quantification theory, and theory of identity) from the other (metaphysical) framework principles—a convention adopted by Kant, among others, but not by Aristotle. Examples of explicit categorial frameworks are the systems or rather subsystems of Kant's *Critique of Pure Reason,* of Leibniz's *Monadology,* and of Whitehead's *Process and Reality.* Categorial frameworks are implicit in the commonsense thought of various persons and societies.

That people do in fact think within categorial frameworks is borne out by their use of the defining logical principles (e.g. of classical or intuitionist logic) as standards of logical necessity, possibility, and impossibility, and of the defining metaphysical principles (e.g. of the principles of causality or complementarity as constitutive of events, or the principles of locating phenomena in Newtonian space and time or Riemannian space-time as individuating for events) as standards of a necessity, possibility, or impossibility which—to distinguish them from the logical variety—are sometimes called "material," "natural," or by smiliar names. On the other hand, that the logical and the metaphysical principles defining a categorial framework are standards of necessity only for those who think within it, and that as a matter of fact different frameworks have been employed by different persons and societies, are cogent reasons for rejecting the doctrine that a particular categorial framework is absolute. Since, however, the possibility of alternative logics has only very recently been made the subject of study by logicians and philosophers, the possibility of frameworks which differ in their defining logical principles is still a controversial issue.

A person's adherence to a categorial framework may be more or less explicit, more or less definite, and more or less confident. His confidence in it may vary from unquestioning acceptance to uncertain wavering between its continued acceptance and its rejection in favor of an alternative framework which for some reason or another may prove more attractive to him. It seems likely that the notion of a categorial framework would be a useful piece of theoretical equipment for the historian of ideas, since he is particularly concerned with intellectual changes that are both fundamental and sufficiently widespread to be of historical rather than merely biographical interest. For example, the transitions from Aristotelianism to what might be called "Kantianism," and from Kantianism to a new ontology inspired by quantum mechanics, could be described as changes of categorial frameworks rather than as changes of paradigms or other equally ill-defined structures. In what follows, however, we are concerned with other applications of the notion of a categorial framework.[1]

2. On the Difference between Primary and Secondary Logical Principles and Their Relation to Metaphysical Principles

Since a person's logical principles are in the widest sense transcategorial in that they govern his thinking about any particular and any characteristic whatsoever, a person's logic and his metaphysics must correspond to each other. Although this correspondence has been often, and on the whole correctly, emphasized, its formulation needs more care than is usually given to it if error and confusion are to be avoided. For this purpose it is useful to consider some prima facie examples of the correspondence. Among the most frequently mentioned are the correspondence between Aristotle's logic (according to which every elementary proposition has one subject and one predicate) and his metaphysics (especially the doctrine that as far as their "reality" is concerned, relations are farthest removed from substances and less real than either qualities or quantities); and the correspondence between Leibniz's logic (according to which the Aristotelian analysis of elementary propositions has to be supplemented by the assumption that every predicate is contained in its subject) and Leibniz's metaphysics (especially the doctrine that all the attributes of a substance are inseparable from it).

Other increasingly controversial examples are a correspondence between the early Wittgenstein's truth-functional logic (in spite of its illegitimate extension to quantification theory) and his metaphysical theory of logical atomism; and a correspondence between the later Wittgenstein's version of a constructivist logic (in spite of its never having been worked out clearly) and his metaphysical doctrine of "forms of life" (which also remains rather obscure). One might also argue that by incorporating certain assumptions of Darwin's theory of evolution into one's metaphysics, one commits oneself to a logic which acknowledges not only attributes which are extensionally definite but also extensionally indefinite ones; and that by incorporating certain assumptions of quantum mechanics into one's metaphysics, one commits oneself to a logic which does not contain the principle of excluded middle.

The relationship between a person's logic and his metaphysics may, however, be more complex. He may in particular vacillate between two competing sets of logical or metaphysical principles or the employment of two logical systems one of which has to function as primary and the other as secondary, that is to say, as subordinated to, and capable of interpretation in, the primary system. An important early example of a philosopher who clearly made this distinction was Leibniz, whose primary logic was a version of the subject–predicate logic, and who developed a logic of relations as a secondary logic in order to be able to reason efficiently about mathematical and other relations which he regarded as mere *entia rationis*, referred to by syncategorematic terms only. More recent examples of the use of one logic as primary and another as secondary are the use by some intuitionists of classical logic (intuitionistically interpreted) for heuristic purposes, e.g. for providing nonconstructive proofs intended as pointers towards possible constructive ones; or the use by classical logicians of intuitionist logic (classically interpreted) for practical purposes, e.g. for communicating their results to intuitionists and thereby improving cooperation with them.

Failure to note this distinction may lead to a familiar type of philosophical confusion, of which it is symptomatic that its victim, when trying to understand a philosopher who makes the distinction, accuses him of being confused or even deliberately confusing. Both these accusations are levelled by Russell against Leibniz who, according to Russell, should have fitted his metaphysics to his secondary logic and not–as in fact he consciously, conscientiously, and openly did–to his primary logic.[2] For the purpose of explaining the relation between logic and metaphysics it is thus important to remember that whenever both primary and secondary logical principles are employed by a person, it is only the former which impose transcategorial constraints on his thinking about any object and any characteristic whatsoever and, hence, on the metaphysical principles which define his categorial framework.

3. On Formal and Material Necessity in the Spheres of Factual and Practical Thinking

The logical principles which (partially) define a person's categorial framework determine what in his sphere of factual thinking is logically or formally necessary, impossible or possible. If, for example, classical quantification theory is his primary logic, then a factual statement, say f, is logically necessary if and only if $\vdash_L f$, logically impossible if and only if $\vdash_L \neg f$, and logically possible if and only if $\neg \vdash_L \neg f$. If his primary logic is intuitionist or some other system of logic than the letter L under the symbol '$\vdash$' would have to be replaced by I or another appropriate letter. (Where the logic referred to is not in doubt or irrelevant in the context, the symbol '$\vdash$' will be used without subscript. The use of the same symbols–e.g., '$\neg$'–for logical and metalogical notions should be harmless in the contexts in which they will occur here.) Deducibility of g from f will be expressed in the usual manner by $\vdash f \rightarrow g$ or $f \vdash g$.

So long as a person adheres to a certain categorial framework (e.g., the Kantian), he will regard the metaphysical principles defining it (e.g., the three analogies of experience) as being not only true but necessary in a sense of "necessity" which will be called "material." If the conjunction of the metaphysical principles defining the framework is, say, F, and the logical principles defining it are the principles of the classical logic L, we might symbolize the material necessity of a statement f with respect to F by $\Box_F$ f and define

(1) $\Box_F f \underset{D}{=} f \wedge .(F \vdash_L f)$.

We might similarly define material possibility with respect to F by

(2a) $\Diamond_F f \underset{D}{=} \neg \Box_F \neg f,$

or more explicitly by

(2b) $\Diamond_F f \underset{D}{=} f \vee \neg(F \vdash_L \neg f),$

and material impossibility with respect to F by

(3) $\neg \Diamond_F f \underset{D}{=} \Box_F \neg f.$

It is of interest to note that if one assumes that L is the classical logic and that F_0 is a particular (constant) conjunction of metaphysical framework principles, Lewis's modal system S_5 is an appropriate formalization of reasoning about material necessities, possibilities, and impossibilities (with respect to F_0). Corresponding to the necessary or strict implication of that modal system we might define

(4) $f \prec_F g \underset{D}{=} \neg \Diamond_F (f \wedge \neg g).$

Some philosophers are in the bad habit of using the notion of coherence as if it were an absolute and well-defined notion. It is neither. But acceptance of the proposal that "coherence" be understood as an elliptical version of "coherence with respect to a certain F" and analysed as '$\Diamond_F f$' would turn it into a well-defined, relative notion.

Within a person's system of practical attitudes (his attitudes toward what can, consistently with his beliefs about the course of nature, be brought about by his conduct) various conflicts may arise on which a distinction between formal and material necessity, possibility and impossibility may also be based. If, for the sake of simplicity, one trichotomizes one's practical preferences into practical proattitudes, practical antiattitudes, and attitudes of practical indifference, and if one acknowledges the *commonly neglected* stratification of practical attitudes (i.e., the possibility and occurrence of practical attitudes towards one's practical attitudes), we may distinguish between the following species of conflict between practical attitudes. (a) Opposition: Two practical attitudes are opposed if and only if their joint implementation is impossible. Two opposed practical attitudes may but need not be complementary, in the sense that the nonimplementation of at least one of them is impossible. (The impossibility may be logical or material in the sense in which these notions apply, as has just been explained, in the sphere of factual thinking.) (b) Discordance: Two practical attitudes are discordant if and only if one of them is an antiattitude toward the other (e.g., a person's practical anti-attitude toward his proattitude toward smoking). (c) Incongruity: Two attitudes are incongruous if each of them is a proattitude toward a different lower-level attitude, and the two lower-level attitudes are opposed to each other; or, if each of them is an anti-attitude toward a different lower-level attitude and the two lower-level attitudes are complementary to each other.[4]

While it is important to distinguish between these types of practical conflict, we may for our present purpose combine them and define two practical attitudes (or two statements expressing them) as practically inconsistent if and only if they are either opposed to, discordant with, or incongruous with each other. Writing *a, b,* etc. to express that a person *S* has a certain distinct practical attitude, and writing $\neg a$ to express that he has an attitude that is practically inconsistent with the attitude expressed by *a* (so that if *a* expresses his having a proattitude toward smoking, then $\neg a$ expresses his having either an antiattitude or an attitude of indifference toward smoking), and writing '$\diamond a$' for 'the practical attitude expressed by *a* cannot be decomposed into a set of practical attitudes of which at least two are practically inconsistent', we can define a notion of formal necessity in the practical sphere by

(5) $\boxdot\, a \underset{D}{=} \neg \diamond \neg a,$

and a notion of practical implication by

(6) $a \Rightarrow b \underset{D}{=} \neg \diamond \ (a \wedge \neg b).$

It may be worth emphasizing again that the notion of material necessity in the factual sphere enters the notion of practical opposition and incongruence and, hence, of formal necessity in the practical sphere.

Just as a person's factual thinking is constrained by formal principles, giving rise to formal necessity, and by metaphysical principles, giving rise to material necessity, so a person's practical thinking is constrained by formal principles, giving rise to formal necessity in the practical sphere, and by principles of practical rationality, giving rise to material necessity in the practical sphere. In this sphere the formal principles would be systematically represented by some "logic of preference" of the kind used, e.g., in decision theory, but taking account of the stratification of practical attitudes. The notions of pro-, anti-, and indifferent practical attitudes, as well as the notions of formal necessity (in the practical sphere) and of a principle of practical rationality, are all definable in this logic—a technical task which is fairly straightforward.

Examples of principles of practical rationality are various principles of utilitarian ethics, Kant's categorial imperative, and various principles of justice. A person's principle of practical rationality expresses a practical attitude which, roughly speaking, "governs" all his practical attitudes without being governed by any of them. While the detailed analysis of these practical principles is not needed for our present purpose,[6] it seems clear that a person accepting such a principle will regard it as not only morally acceptable but necessary in a sense of practical necessity which, in order to distinguish it from formal necessity, is appropriately called "material." If we symbolize a person's notion of moral acceptability by *Mor* and the conjunction of his principles of practical rationality by *P*, we might symbolize material necessity with respect to *P* by $\boxdot_P a$ and define

(7) $\boxdot_P a \underset{D}{=} Mor\,(a) \wedge (P \Rightarrow a).$

We can similarly define material possibility or practical coherence

(8) $\Diamond_P\, a \underset{D}{=} Mor\,(a) \vee \neg (P \Rightarrow \neg a).$

The analogy between (7) and (8) on the one hand and (1) and (2) on the other, is obvious, even though it is much more common in ethics to identify moral acceptability with deducibility from principles of practical rationality than it is in epistemology to identify truth with deducibility from metaphysical principles. The reason for discussing, although in a necessarily sketchy manner, formal and material necessity and related concepts not only in the sphere of factual but also in the sphere of practical thinking is that, contrary to first impressions, both groups of concepts play an important role in metaphysical arguments, to which we now turn.

II

4. Metaphysical Arguments Based on Formal and Material Necessities in Factual and Practical Thinking

The distinctions which have just been drawn between formal and material necessity in the spheres of a person's factual and practical thinking make it clear that his acknowledgment of each distinct kind of necessity and the reasoning based on it imply the acceptance of a primary logic. He needs it if he is to determine whether in the factual sphere one of the requirements for a proposition's material necessity is satisfied, namely, its logical deducibility from the principles which define his categorial framework. He similarly, though perhaps less obviously, needs his primary logic if he is to determine whether in the practical sphere one of the requirements for an attitude's formal or material necessity is satisfied, namely the practical realizability of the attitude. (To believe that an attitude is practically realizable is to believe that its realization is *logically* compatible with one's beliefs about the course of nature.)

To emphasize such tacit or implicit use of a person's primary logic in his reliance on nonlogical necessities is not to deny—as is done by various positivists and other antimetaphysical metaphysicians—that it may not also be used in the explicit deduction of clearly expressed metaphysical conclusions from clearly expressed metaphysical premises. Thus, to give one example (and to risk belaboring the obvious), one may mention Leibniz's argument that if all substances are monads, then "the action of one substance on another" is "neither the emission nor the transplantation of an entity," as is believed by "the vulgar" (who include not only most of the ancient and scholastic philosophers but also Descartes).[7] The argument is purely deductive. In particular, it does not rely on the assumption of allegedly indispensable framework principles, whose special status in the sphere of a person's factual thinking invests some nonlogical connections and arguments with "material" necessity.

As against this, a proof, or attempted proof, of a proposition's material necessity in the sphere of a person's factual thinking requires, beyond its deduction from certain premises, an acknowledgment of the indispensability of these premises. The indispensability of these premises in the person's factual sphere—their defining his categorial framework—may be asserted as a simple fact, or as capable of explanation by a philosophical or other theory. An example of the more elaborate strategy is Kant's proof of the material (synthetic a priori) necessity of the so-called "law of the continuity of all alteration." It consists in (a) the deduction of the law from the assumption of the continuous structure of time and temporal phenomena and the principle of causality, (b) the assertion of the obvious truth and indispensability of the two premises, and (c) an explanation of their indispensability by his theory of time as the a priori condition of sense experience and of the understanding (and its principles, including the principle of causality), as the a priori condition of the knowledge of objects.[8] An example of the simpler strategy is Hume's argument that a deterministic conception of human liberty is materially necessary. For he simply asserts, by means of some rhetorical questions, that history, politics, and any foundation of morals would be impossible without the assumption of unrestricted determinism, and he clearly implies that without it one could not think intelligently or intelligibly about human actions.[9]

Metaphysical arguments based on the material necessity of practical attitudes in the sense of their being morally acceptable and practically implied by principles of practical rationality (see (7) above) are most readily identified in Western theology and theologically inspired metaphysical systems. In such systems a principle of absolute perfection is ascribed to God and used to justify the acceptance of metaphysical principles and principles of practical rationality that are considered appropriate to finite human beings. Thus, according to Leibniz's metaphysics, it is the will of God that the world should contain "the greatest possible variety together with the greatest possible order" (*Monadology* §58). This principle of divine optimization, which is explained in the *Theodicy* and other writings, practically implies an antiattitude toward Huxley's merely possible Brave New World, as well as a proattitude toward the existence of the actual world in spite of its apparent imperfections. Since all its features are materially implied by the principle of perfection and—being compatible with God's infinite goodness—morally acceptable, they are materially necessary in God's practical sphere or in the sphere of divine creation. From the point of view of human beings, who try to understand it according to their lights, this material necessity is, strictly speaking, not practical but factual.

Yet metaphysical arguments can also be based directly on material necessities in the practical sphere, without a theological detour. The occasion for such an argument might arise for a person choosing between two mutually exclusive metaphysical assumptions, each of which would make his categorial framework (say, the constitutive attributes of a certain maximal kind) more determinate, provided that the two assumptions are each logically compatible with his present metaphysical principles, that neither is by any standards more plausible than the other, and that the adoption of one of them would be immoral. Bellarmine's reasoning against the adoption of Galileo's cosmology (as a correction of Christian doctrine rather than as a mere working hypothesis) has been interpreted by some historians in this way. While metaphysical arguments that are at least to some extent based on practical implications, and the corresponding material necessities, are particularly attractive to pragmatist philosophers, they are also bound to remain among the favorite arguments of men of action, engaged in the persecution or forcible "reeducation" of religious or political dissenters and in attempts at justifying this activity.

In separating, as has been done here, concepts and arguments belonging to the factual from concepts and arguments belonging to the practical sphere, one must not obscure their intimate interdependence. The connection, which enters the very notion of a practical attitude, is perhaps most obvious in the case of a belief the holding of which essentially involves a proattitude toward certain kinds of conduct—essentially, in the sense that the absence of the belief in a person would *eo ipso* involve the absence of the attitude. Quite apart from such "instrumental beliefs" (e.g., that a certain drug is poisonous or that prayer may bring about beneficial miracles), a little reflection shows that having practical preferences between any types of option presupposes some classificatory scheme and thus beliefs about the applicability and application of certain attributes to particulars. Conversely, it shows that the choice between different classificatory schemes depends on practical preferences, e.g., for schemes which are more simply or efficiently applicable. A deeper inquiry into the interdependence of factual and practical thinking might well start with a careful examination of controversies between constructivist and classical logicians as to whether classical or constructive logic is primary, or study of controversies between metaphysicians about the adequacy or inadequacy of the Kantian thesis of the "primacy of practical over pure reason." This, however, is not the place for such an inquiry or even for indicating the lessons that might be learned from it.

5. Metaphysical Arguments Based on Plausibility and Thereby in Part on Formal and Material Possibilities in the Spheres of Factual and Practical Thinking

A person's formal principles in the factual and practical sphere (e.g., the principle of excluded middle and the requirement of the nonopposition, concordance, and congruity of practical attitudes), the metaphysical principles defining his categorial framework (e.g., a Kantian's principle of causality), and his principles of practical rationality (e.g., a Kantian's supreme practical principle never to use another person as a means only), determine not only what for him in the factual and practical sphere is formally and materially necessary but also what for him in these spheres is formally or materially possible. (See the definitions of §3.) In other words, no connection—or argument based on it—is considered plausible by a person who considers it formally or materially impossible in the factual or practical sense of these terms. And no plausibility argument is correct for a person unless it is in both these senses formally and materially possible.

In order to understand the function of this requirement in metaphysical plausibility arguments and their general structure, it is convenient to distinguish among the following elements or stages in the strategy of such arguments:
(i) An exhibition or proposal of a metaphysical principle, i.e., the statement that the principle has been accepted by a person or group of persons or the proposal that it be so accepted in addition to already accepted principles or in place of one of them (e.g., the principle of causality or one of its contraries such as the principle of complementarity);
(ii) if the metaphysical principle is exhibited as accepted rather than proposed for acceptance, an empirical demonstration of the acceptance;
(iii) a factual possibility demonstration, i.e., a demonstration that

the metaphysical principle is in the factual sense formally possible (logically consistent with the addressees' primary logic) and materially possible (logically consistent with the addressees' unreplaced nonlogical framework principles);
(iv) if the metaphysical principle expresses an instrumental belief, a practical possibility demonstration, i.e., a demonstration that the proattitude involved in the belief (e.g., toward various lines of scientific research) is in the practical sense formally possible (not decomposable into opposed, discordant, or incongruous attitudes) and materially possible (not practically inconsistent with practical attitudes expressing the addressees' supreme principles of practical rationality); and lastly,
(v) a plausibility demonstration in the narrow sense of the term, i.e., an attempt at justifying the exhibited or proposed principles beyond providing the factual possibility demonstration or, where required, the factual and practical possibility demonstrations.

No comment is needed on the exhibition or proposal of a metaphysical principle, and hardly any on the empirical demonstration or on the factual and practical possibility demonstration. The former does not differ in principle from the manner in which the presence of beliefs and attitudes in oneself or others is established by historians, anthropologists, grammarians, or interested laymen. The latter consists in the application to the compared beliefs and attitudes of the relevant concepts of possibility. As to the demonstration that an exhibited or proposed metaphysical principle is plausible, it may take many forms of which the following are fairly common.

One, which might be called the "no option" argument, applies in the nature of the case only to exhibited principles. In it the person putting it forward simply points out that the principle is in fact employed by him and the addressees of the argument, if any, in thinking about their experience or a region of it, and that, at least for the time being, no feasible alternative is conceivable to them. The refutation of the argument consists in showing that such an alternative is, or has become, conceivable to them. For example, the "no option" argument that post-Galilean scientific and commonsense thinking about the physical world cannot, as a matter of fact, dispense with the principles of causality and continuity was unanswerable until 1925, when it could be refuted by pointing to the existence of quantum mechanics.

Another similar, but more complex, plausibility demonstration, might be called the "*faute de mieux*" argument. In it the person putting it forward points out that the exhibited or proposed metaphysical principle (set of principles or even categorial framework) is of all those that are conceivable to him and the addressees of his argument the best, that is to say the best by a standard on which he and they agree. The degree of complexity of this kind of argument depends to a large extent on the nature of the agreed standard. It will on the whole be smaller if the agreed standard is applied directly to the compared principles (e.g., if it is a straightforward test of simplicity after the fashion of Ockham's razor), and greater if it is applied indirectly by considering their employment in other fields (e.g., in scientific thinking by comparing, say, the predictive power of theories constructed in accordance with the compared principles such as the principle of causality and the principle of complementarity). In the latter case the plausibility demonstration has to rely on inductive arguments, and this raises the general problem of induction.[10]

Lastly, one should mention a form of plausibility demonstration based on what might be called "the aesthetic appeal of free speculation." It consists in confronting oneself and others with a metaphysical principle or system that can be shown to be neither factually nor practically impossible or implausible, and whose contemplation—like that of some beautiful objects of nature or art—produces a more powerful impression of aesthetic and intellectual satisfaction than that produced by other metaphysical principles or systems with which it is compared. Such an appeal can be claimed not only for the metaphysical systems of visionaries like Cusanus or Spinoza but also for metaphysical principles whose material possibility and plausibility in the factual and practical sphere are in part controlled by the predictive and technological aims of modern science. An example is Einstein's often quoted refusal to believe in a "dice-playing God."

6. On Some Characteristic Fallacies of Metaphysical Thinking

The various types of necessary connection—formal and material, factual and practical—which have been discerned and shown to be the basis of sound metaphysical arguments, may also through misunderstanding and confusion become a source of metaphysical arguments which merely appear to be sound. Thus metaphysical thinking is clearly not immune from the ordinary logical fallacies (affirming the consequent, begging the question, etc.) associated with ordinary logical implication (say, $f \underset{L}{\vdash} g$ or $\underset{L}{\vdash} f \rightarrow g$). Again, insofar as the various "special" implications used in metaphysical thinking (e.g., $f \underset{F}{\dashv} g$ and $a \Rightarrow b$ as defined by (4) and (6) of §3) are structurally similar to ordinary logical implication, the associated fallacies are also similar.

More important, because more insidious, are the fallacies that consist in confusing a properly established weak metaphysical claim with a stronger metaphysical claim which has not been established—a confusion which, if it were deliberate, would conform to the sophistical art of making the weaker statement the stronger (τὸν ἥττω λόγον κρείττω ποιεῖν).[11] The following is a sequence of increasingly stronger claims, each of which tends to be, and has been, confused with its successor in such a way that a satisfactory demonstration of the weaker claim is *eo ipso* regarded as demonstrating the stronger. The members of this sequence are (i) finding that the negation of a metaphysical principle, say *f*, appears to be odd to the person or persons considering it (schematically, $Odd \neg f$); (ii) regarding *f* as materially necessary with respect to a particular categorial framework F_o, say the framework of the person or persons considering it (schematically, $\Box_{F_o} f$); (iii) regarding *f* as materially necessary not only with respect to F_o but also with respect to some other frameworks, say $F_1 \ldots . F_n$, which are conceivable to the person or persons considering *f* (schematically, $\Box_{F_o \wedge F_1 \wedge \ldots \wedge F_n} f$); (iv) regarding *f* as materially necessary not only with respect to frameworks which are for the time being conceivable to the person or persons considering *f*, but with respect to any possible frameworks (schematically $\Box_{F_o \wedge F_1 \wedge \ldots \wedge F_n \wedge \ldots} f$); and lastly, (v) regarding *f* as absolutely necessary in some sense not covered by the senses of logical or nonlogical necessity which have so far been considered.

The reason why the odd appearance of $\neg f$ does not imply the material necessity of *f* with respect to the framework F_o of a person considering $\neg f$ is that the odd appearance to him of $\neg f$ may also be due to other circumstances. He may, for example, only very recently have undergone a metaphysical conversion to F_o and be the kind of person whose linguistic and other intellectual habits die hard, or he may be wavering between two competing categorial frameworks. That the material necessity of *f* with respect to one framework, say F_o, does not imply the material necessity of *f* with respect to F_o and others, say $F_1 \ldots . F_n$, is obvious, once the alleged implication is clearly formulated. The same applies to the statement that the material necessity of *f* with respect to the specified frameworks $F_0 \ldots . F_n$ implies its necessity with respect to all possible frameworks $F_0 \ldots . . F_n \ldots . . .$, whether or not they are specified or specifiable. A further defect of this invalid implication is the obscurity of its consequent. The obscurity becomes overwhelming for the last claim in our sequence, which contains the statement that *f* is in some nonlogical, absolute, and unexplained sense necessary and affects any alleged logical implication in which this statement occurs as consequent.

There are many other types of fallacy characteristic of metaphysical thinking. Among them are the so-called transcendental arguments to the effect that a certain categorial framework (say the Kantian or that found in Wittgenstein's *Tractatus*) is unique in being true and absolutely necessary. A defect shared by all these arguments is that the required uniqueness proof is either not provided or consists in an illegitimate transition from statements about specified categorial frameworks to statements about all possible ones. In a systematic exploration of further fallacies of metaphysical reasoning, one might note that in the factual sphere the temptation to replace material necessity with respect to one or more categorial frameworks by an obscure, absolute notion of necessity extends also to material possibility and thus to various notions of plausibility. Again, one might construct a sequence of increasingly stronger practical claims and show how they, like the corresponding factual claims, can be, and often are, confused in metaphysical reasoning.

The examination of metaphysical arguments reveals, as does the examination of mathematical, scientific, and legal arguments, a great variety of systems which are properly described as similar in structure and function but different in content. It also raises the question whether or not there are limits to the divergence of these systems, determined for example by a core of principles common to all of them. But just as the corresponding questions about mathematical, scientific, and legal systems are not themselves mathematical, scientific, or legal, so is this question about metaphysics not metaphysical. It is an empirical or, more precisely, an anthropological question—even though as matters stand at present in the republic of learning, philosophers might be better equipped to answer it than anthropologists.

3: RUTH BARCAN MARCUS

Does the Principle of Substitutivity Rest on a Mistake?

1.

In the paper "Attribute and Class,"[1] Frederic Fitch suggests that "if entities X and Y have been identified with each other, it seems reasonable to suppose that the *names* of X and Y should be everywhere intersubstitutable where they are being used as names." The principle of substitution is here proposed as a reasonable supposition. Some philosophers are less tentative in their acceptance, as is Quine when he says: "One of the fundamental principles *governing* identity is that of substitutivity. . . . It provides that given a true statement of identity one of the two terms may be substituted for the other in any true statement and the result will be true."[2] Others are more tentative, going so far as to claim that however reasonable a supposition the principle of substitutivity may appear to be, it is simply false. They point, for example, to the sentences

(1) Georgione is so called because of his size,

and

(2) Barbarelli is so called because of his size.

Here 'Georgione' and 'Barbarelli' are names being used as names, and though 'Barbarelli is Georgione' is true, substitution converts a truth to a falsehood.

Recently Cartwright,[3] who rejects the principle of substitutivity on the basis of such counterexamples, has set about to explain why the belief persists that it is a governing principle. The error, he claims, is that it has been confounded with another principle, the principle of indiscernibility (or as he prefers to describe it, the principle of identity). The latter, however, is not concerned with expressions and substitutions but rather with the properties things have: It says

(Ind) If $a = b$ then every property of a is a property of b.

He goes on to argue that although (1) and (2) provide a counterexample to substitutivity, it does not follow from this that such counterexamples are also counterexamples to the principle of indiscernibility. For (1) and (2) to count as a counterexample to the principle of indiscernibility, there must be some property that Georgione has and Barbarelli lacks. But if from (1) we form the associated predicate

(3) x is so called because of his size,

we find it does not determine a property. The quasi-predicate 'being so called because of his size' does not specify a function that takes us from individuals to propositions. Other attempts to specify a property attributed to Georgione in (1) and not shared by Barbarelli are shown to be incoherent. If, on the other hand, the property attributed to Georgione is that of being called 'Georgione' because of his size, this *is* a property which Barbarelli has, and we do not have a counterexample to (Ind). Cartwright goes on to examine other cases of substitution failure such as puzzles about nine and the number of planets, and argues that here too failures of substitutivity are not at the same time failures of (Ind).

But is the connection between the principles of identity and substitutivity a case of mistaken identity? Some philosophers have clearly distinguished them from each other[4]—but have claimed as well that they are both versions of Liebniz's Law: If a and b are identical, then whatever is true of a is true of b. One version, (Ind), is perceived as being in the material mode, while substitutivity is perceived as being in the formal mode. Indeed, in formal languages the two principles are intimately connected. Substitutivity may be taken as a rule of derivation in first-order logic with identity. In the absence of quantification over properties, the principle of indiscernibility comes to the set of those valid sentences that are the associated conditionals of the rule of substitutivity for each of the predicates of the theory. Alternatively, (Ind) may be taken as definitive of identity in second-order logic, and the principle of substitutivity falls out metatheoretically. Anyone who claims that substitutivity doesn't govern identity but that indiscernibility does must wonder why this is not so in formal languages. If translation into predicate logic is supposed to display the logical form of at least some important fragments of ordinary discourse, how can substitutivity rest on a mistake?

What Cartwright has noted is not peculiar to the principle of substitutivity. Apparent counterexamples can be found in natural language for virtually all logical principles where surface grammar is untransformed or the resolution of ambiguities is implicit and resolvable only by attention to the context of use. Does the truth of 'Socrates is sitting' and 'Socrates is not sitting' provide a counterexample to the principle of noncontradiction? Homonyms and the absence in ordinary speech of a rule, one name on thing, generate false sentences of the form '$x = x$'. Do they count as counterexamples to a logical law?

If we are to deny (as we do) that belief in the principle of substitutivity is a straightforward case of mistaken identity, the question of the principle's logical status remains. Prima facie, substitutivity is not an empirical principle about language use. Is it a regulative or normative principle? An answer here depends on how one explains, or explains away, apparent failures of the principle in ordinary English, even when the latter is supplemented by some of the apparatus of formal logic.

2.

Central to an understanding of the role of a principle like that of substitutivity is the belief that there are different ways of saying the same thing and that some ways are logically preferable in that they reveal logical form. Criteria for judging a way of saying as being preferable include that it resolves ambiguities and dispels puzzles or antinomies. Further desiderata may be more controversial. For those of a non-Meinongian temper, reduction of ontology might be a criterion. Frege states (or perhaps overstates) the case as follows:

> We must not fail to recognize that . . . the same thought may be variously expressed. . . . It is possible for one sentence to give no more and no less information than another; and for all the multiplicity of languages, mankind has a common stock of thoughts. If all transformation of the expression were forbidden on the plea that this would alter the content as well, logic would

> simply be crippled; for the task of logic can hardly be performed without trying to recognize the thought in its manifold guises.[5]

One of the tasks of logic, then, is to select from among all the sentences that express the same thought those which express it better. Whether there is a best way, "*the* logical form," supposes, as did Wittgenstein in the *Tractatus,* that "there is one and only one complete analysis of the proposition." That supposition remains controversial. The point of this paper is simply that a principle like substitutivity is to be tested with respect to statements in logical form.

But the pursuit of logical form is multiply complicated. Logicians, until recently preoccupied with the task of selecting from among sentences that express the same proposition the ones (one) that better (best) express it, have neglected the obvious fact of usage that behind the same sentence (type) may lie different thoughts. Russell noted the ambiguity of sentences like 'George IV wonders whether Scott is the author of Waverly' and proposed alternative analyses for each interpretation. Similarly, we are familiar with amphibolies like 'Everyone loves someone'. But for a host of sentences the same sentence-form may express different propositions, and the ambiguity is not to be explained in terms of structural factors. These sentences are ambiguous because a determination of the objects being talked about on a given occasion requires that we have knowledge of the wider context in which the sentence has been uttered on each occasion of its utterance. Cartwright, for example, in noting the ambiguity of (3), takes it as obvious that (1) is unambiguous in expressing a proposition. But in the presence of a personal pronoun, that clearly is not necessarily so. Suppose (1) had been uttered in response to 'Shorty is so called because of his size'. In such a setting (1) might express that Georgione is called 'Shorty' because of Shorty's size, that Georgione is called 'Shorty' because of his own (Georgione's) size, or that Georgione is called 'Georgione' because of his own (Georgione's) size, etc.

3.

What is a correct formulation of the principle of substitutivity? Cartwright takes it to be the following:

(Sub) For all expressions α and β, $\ulcorner \alpha = \beta \urcorner$ expresses a true

> proposition if and only if, for all sentences S and S′, if S′ is like S save for containing an occurrence of β where S contains an occurrence of α, then S expresses a true proposition only if S′ does also.[6]

An obvious inadequacy of (Sub), if '=' is to be interpreted as identity, is a failure to specify a substitution class for α and β. In the absence of such a specification, the right side of an instance of the biconditional (Sub) could be true and the left side simply meaningless, e.g., the sentence 'just in case = if and only if' makes no sense although substitution of 'just in case' for 'if and only if' may be truth-preserving. The preservation of meaning of an identity sentence requires in the least that the grammatical category be that of singular terms construed broadly as including demonstratives, pronouns, names, and descriptions, but a syntactic criteria is insufficient. A further condition on a meaningful identity sentence is that terms which flank the identity sign refer. Whether or not a term refers, however, is not a syntactic matter. There are, furthermore, different ways of referring, and not all of them are verbally realized. The mere utterance 'this is identical with this', without accompanying gestures or clues is not a meaningful identity sentence, yet since they are homonymous expressions, truth is preserved by substitution of 'this' for 'this'. With accompanying clues, an identity sign flanked by demonstratives or pronouns is on a given occasion of use meaningful and often true, but on different occasions the same locution and gestures may refer to different objects. For such cases, although the identity sentence may be meaningful, it is not reasonable to suppose that (Sub) is a governing principle.

The remaining candidates for the substitution class in (Sub) are definite descriptions and (proper) names, but there are grounds for eliminating the former. We are speaking here of a definite description being used in the ordinary way, namely, to pick out an object (if any, and then only one) that satisfies the description. A digression is warranted at this juncture. Much has been made recently of the fact that a description, even an erroneous one, may serve on occasion to (purely) refer to an object.[7] That may be pragmatically interesting, but such uses are more akin to the use of demonstratives. 'The woman over there drinking a martini' may serve on a particular occasion to tag a particular man drinking champagne, but if its use is idiosyncratic and transient, it is like a discardable tag, for it does not enter into the common language. Occasionally such ways of referring become entrenched, and a measure of such entrenchment is the conversion of a faulty description to a proper name. Ordinary English has a way of marking the change by using capitals. 'The Evening Star' serves as an example. It is curious that so simple a typographical device for separating the proper name-like use of a description from its predicative use has never been employed in formal analysis. How directly the apparent contradiction in 'The evening star is not a star' is dispelled by 'The Evening Star is not a star'.

Unlike the exclusion of demonstratives and pronouns, definite descriptions are excluded from the substitution class in (Sub) on the ground that the logical form of an identity sentence flanked by a description (used descriptively) is not given by '$x = y$'. Supporting this analysis are the reasons Russell laid out so meticulously. An identity sentence, true or false, presumes reference, and descriptions need not refer. Further reasons are entailed by the nature of the identity relation itself, which is the relation between an object and an object just in case it is the same object. That an object is the same as itself is not a contingent matter, but that one object uniquely satisfies two sets of properties may well be. We are not propounding the theory of descriptions as *the* analysis of all 'the' phrases. There are other and perhaps better ones particularly those which deal with abstract objects such as numbers. But whatever the analysis, it should display the predicative role of descriptions as ordinarily used, which is not revealed in the surface grammar of an identity sentence flanked by a definite description.

To specify the substitution class as the class of proper names is not merely to specify a syntactic category. To count as a proper name, the expression must refer without being tied to any particular characterization of the object. It requires a naming episode and conditions for the name to enter into the common language as do other words in common speech. Proper names have a logically irreducible use. They permit us to entertain a separation in language of the object under discussion from its properties. How a word, a name, can come to have such a use has been articulated by Donnellan, Kripke and others.[8] Geach puts it succinctly as follows:

For the use of a word as a proper name there must in the first

> instance be someone acquainted with the object named. But language is an institution, . . . and the use of a given name for a given object, . . . like other features of language, can be handed on from one generation to another; . . . Plato knew Socrates, and Aristotle knew Plato, and Theophrastus knew Aristotle, and so on in apostolic succession down to our own times; that is why we can legitimately use 'Socrates' as a name the way we do. It is not our knowledge of this chain that validates our use, but the existence of such a chain. When a serious doubt arises . . . whether the chain does reach right up to the object named, our right to use the name is questionable, just on that account. But a right may obtain even when it is open to question.[9]

In addition to a specification of the substitution class as proper names, what further condition must be placed on (Sub)? As noted above, the counterexample posed by (1) and (2) is not to be accounted for by a failure to use proper names or by an incomplete analysis of the identity sentence 'Georgione = Barbarelli'. At this juncture Cartwright's analysis is suggestive of additional constraints to place on (Sub) if it is to reflect Leibniz's law, that if *a* and *b* are identical, whatever is true of *a* is true of *b*. Cartwright notes that what is true of Georgione, i.e., being so called because of his size, is not "detachable." Replacing the names by place markers doesn't "capture," so to speak, the predicate. When a sentence contains proper names properly used, properties or relations are ascribed to the objects named, and a representation of such a sentence in logical form should articulate the partition. Displaying form is not like mapping reality. Nor is it very helpful to speak of logical form as "displaying the form of the facts." It is rather a way of saying what we want to say, unambiguously, minimizing a dependence on non-verbal context for interpretation and preserving principles of logic and inference. The failure of a sentence to support substitution of the names of identified objects is a measure of the deficiency of the sentence with respect to logical form. The principle of substitutivity may therefore be perceived as both true and regulative. Let us restate it as follows:

(Sub′) For all proper names α and β (indexed to preserve univocality), $\ulcorner \alpha = \beta \urcorner$ expresses a true proposition just in case for all sentences P, S, and S′, if S is a restatement of P in logical form, and if S′ is like S save for containing an occurrence of β where S contains an occurrence of α, then S expresses a true proposition only if S′ does also.

In proposing (Sub′) we are not also supposing that there is one and only one transformation of a sentence into logical form, for as we noted above, inextricable from all such proposed transformations is the ontological perspective from which they are recommended. I'm thinking here, for example, of the various solutions which have been offered for the puzzle about 9 and the number of planets. An analysis of

(4) 9 is the number of planets

will depend partially on how one views numbers. Are they sets, properties, or first-order individuals? How one interprets the 'is' of (4) will be guided by such commitments. The familiar solution to the puzzle about substitution failure in modal contexts of '9' and 'the number of planets' takes it that while '9' names an individual, the surface grammar identity sign is flanked by a description in '9 = the number of planets' and, (Sub′) therefore, does not apply. Analysis in accordance with the theory of descriptions dispels the substitution failure. But there are alternative analyses. One of them takes it that 9 is a property of sets, and (4) is a deviant way of saying that there are 9 planets. On the latter analysis, although it is necessary that a set with nine members have more than seven members, it does not follow that there are necessarily more than seven planets, since it is a contingency that there are nine of them. The chain of sentences which reflects such an analysis will not generate a substitution failure. But the ultimate ground for choosing between such alternative analyses is not yet resolved. The absence of an account of names which is more adequate to expressions like numerals, weighs against the first alternative, but not decisively.[10]

A belief in the principle of substitutivity is grounded in the belief that the pursuit of logical form is not futile. Proper analysis has in fact dispelled apparent failures of substitution in modal contexts, but others, like epistemological contexts, remain insufficiently analyzed. Nevertheless, as Fitch suggests, it still remains *reasonable* to suppose that the names of identified objects "should be everywhere intersubstitutable where they are being used as names."

4: BAS C. VAN FRAASSEN

Platonism's Pyrrhic Victory[1]

During my acquaintance with Frederic Fitch, I have learned much from him, both through his writings and in discussion about semantic paradoxes and about modal, deontic, and combinatory logic. But I am most grateful to him for the challenge of his ontological position, with which I am in total temperamental disagreement. Temperament is no substitute for arguments, of course, and it rues me to think how much less articulate or well founded than his my ontological position is. In this paper I shall do my best to spell out why I do not feel a need to believe in the existence of abstract entities.

1. The Worlds of Oz and Id

My dialectic strategy will be a bit roundabout. I shall first tell a fable, for the reader to mull over while grappling with the main arguments; in conclusion I shall comment on the fable.

Once upon a time there were two possible worlds, Oz and Id. These worlds were very much alike and, indeed, very much like our world. Specifically, their inhabitants developed exactly the mathematics and mathematical logic we have today. The main differences were two: (a) In Oz, sets really existed, and in Id no abstract entities existed, but (b) in Id, mathematicians and philosophers were almost universally Platonist, while in Oz they refused, almost to a man, to believe that there existed any abstract entities.

They all lived happily ever after.

2. The Strategy of Realism

There are certain phenomena to be taken into account in the philosophy of mathematics. These include the prevalence of language games roughly codified in set theory, classical analysis, and modal logic. The realist gives an account of these phenomena: the axioms of set theory (his favorite set theory) are literally true, he says; there *exist* sets of the sort described. Depending on his ulterior philosophical motives, he may add that certain other abstract entities also exist, or that sets are logical constructs built from other abstract entities (properties or propositions, say). But, on the other hand, he may consider himself moderate and hold that all other abstract entities are built from sets. These sectarian differences we may ignore; they all hold that there exists a family of abstract entities properly called *sets* because they are as described in some set theory sufficiently powerful to serve as a foundation for a sufficiently large body of mathematics and logic.

This is, of course, in its simpler forms the simplest possible hypothesis accounting for the phenomena: the explanation of mathematical theories is that they are true theories in just the way that botany is a true theory. The challenge to anyone not inclined to realism is to provide an alternative account and to subject this account to the criteria of simplicity and explanatory power. In the forties a small band of men valiantly picked up the gauntlet, proclaimed themselves nominalists, set out to "reconstruct" the phenomena nominalistically, and went down to ignominious defeat.

A realist has relatively little work if he so wishes: once he says, say, that Zermelo-Fraenkel set theory is *true,* there already exists a powerful and elegant theory describing his postulated entities, namely, Zermelo-Fraenkel set theory. Any nonrealist is challenged to provide an alternative theory, free of existence-postulates (beyond the acknowledged existence of a finite number of concrete entities) and powerful enough to "reconstruct" this set theory. A Herculean task indeed, I would say, if Hercules' intellect had not lagged so miserably behind his muscles.

Realism wins by default. And that has been the realist's strategy through the ages, to win by default once he has spelled out what his opponent must do to win.

3. The World of Realism

At the turn of the century, the axiomatic method had progressed from construction *more geometrico* to formalization. Proceeding axiomatically had at first consisted of choosing some technical or common terms as primitive and then using these terms plus syncategorematic devices of natural language to formulate one's postulates. Formalization, the *fin de siècle* refinement, eliminated that vestige of reliance on natural language. To intuitionists and their immediate forebears formalization could be at most a means for presenting and communicating mathematical constructions already at hand. But others held the more extreme view that

(a) the formal axiomatic formulation of a mathematical theory captures, in principle, exactly what can be known about the subject matter of that theory;

(b) mathematical concepts are *implicitly defined;* that is, their total significance and meaning is presented through the axiomatic formulation;

(c) an adequate axiomatic theory describes its subject matter *categorically* (that is, uniquely up to isomorphism), and for every mathematical subject there exists, in principle, such a theory.

If today we cannot return to that first fine careless rapture, it must be blamed on a number of metamathematical theorems proved in the half century that followed. I shall state these intuitively, adding precision only where the discussion requires it. The first two are the most famous:

I. (Gödel) A sufficiently strong axiomatic theory is not complete if it is consistent.

II. (Tarski) A sufficiently rich language lacks some means of expression.

If the reader wishes to put these more concretely, he can take "sufficiently strong" and "sufficiently rich" to refer to recursive arithmetic. Hardly less famous is

III. (Löwenheim-Skolem) Any formalized theory has a model which is at most denumerable.

III applies specifically to theories which have axioms saying that there are more than denumerably many things, sets or whatever. Since isomorphism implies equality of cardinality, we have the corollary that no theory about the higher infinities is categorical (if there really are higher infinities). Various philosophers have tried to make it look as if the Löwenheim-Skolem result is innocuous because it involves strained reinterpretations of the primitive terms. But in the proof supplied by Tarski and Vaught it is quite clear that this is not so: any model of a theory will have an at most denumerable submodel of that same theory.[2] This strong form of III was subsequently used by Paul Cohen to derive

IV. (Cohen) The axiom of choice and the continuum hypothesis are independent of the main axioms of set theory.

This theorem is not so general or breathtaking as the first three, but will also appear in my discussion below; a fifth, due to Beth, I shall state later.

My next objective is to consider what reactions to these results are open to the mathematical realist.

I shall divide the realists into the *extreme* (such as Gödel) and the *moderate* (such as Beth). Common to both sects are the following tenets:

(1) The entities purportedly described by mathematical theories exist, and exist independent of any theorizing activity.

(2) These entities are not, and as we now know (see I–III), cannot be adequately (completely or categorically) described by our theories.

(3) Nevertheless, it is possible to refer unambiguously to these mathematical entities, for example, to *the* natural number sequence, or to *the* intended model of classical analysis.

(3) is strongly qualified by (2), in that although we refer to such entities, we do not pretend to have uniquely identifying descriptions for them.

Where extremists and moderates divide is on the adumbration of (3). According to the extremists, the lack of uniquely identifying descriptions does not deprive the word "unambiguously" of any force in (3). We can refer to the number zero or to the natural number sequence and *communicate* to other persons also engaged in mathematics what we are referring to. (There may be gradations in this extremism; perhaps some would say that we can so refer to and communicate about the null set or the class of well-founded sets, though not the number zero. These intrasectarian differences are not important to this discussion.) The moderates, however, feel that there is no way to guarantee that two mathematicians are referring to the same entities. Since they share the same axioms, this need have no practical effects: mathematician *X* accepts mathematician *Y*'s results because those results hold in *X*'s domain of reference, as he can plainly see. And when *Y* begins to articulate postulates that sound strange to *X*, then *X* soon manages to find a model for *Y*'s assertions within his own interpretation.

Beth described the situation graphically in his discussion of Skolem's results on categoricity:

> Set theory has thus at least two models not isomorphic with each other, which we may denote as M_0 and M_1; a model of set theory I shall call for reasons which will soon become clear, a *milky way system.* Let M_0 be the milky way system which we represent to ourselves when we use set theory, or mean to so represent.
>
> If we take a closer look at the categoricity proof for the Dedekind-Peano axiom system (for natural number arithmetic), we find that what is demonstrated is that any two of its models *which belong to the same milky way system* are isomorphic. "The" natural number sequence belongs to the milky way system M_0; the model constructed by Skolem belongs to a different milky way system, say M_1.[3]

And again, a page later:

> In short, the relativism discovered by Skolem in the theory of sets can be described as follows: there are at least two, and likely more, milky way systems each of which provides a complete realization of the complex of logical and mathematical theories, while neither these milky way systems nor the realizations of theories contained in them, are isomorphic.
>
> And yet we imagine that . . . we possess nevertheless an unambiguous intuitive representation of the *structure* of "the" natural number sequence and of "the" Euclidean space. We imagine that our mathematical reasoning intends [has as subject] a specific milky way system M_0, which is intuitively known to us, although this reasoning is equally valid for other milky way systems.[4]

The resolute use of the first-person plural in these passages, published in 1948, is undermined by Beth's amusing, if somewhat cryptic, paper in the Carnap volume. In this paper, which was I think misunderstood by Carnap, Beth imagines a confrontation between Carnap and a second (hypothetical) logician Carnap* who seems to have an alternative milky way system. Beth concludes:

> The above considerations, which are only variants of the Skolem-Löwenheim paradox, suggests strongly that, if arguments as contained in [*The Logical Syntax of Language*] serve a certain purpose, this can only be the case on account of the fact that they are interpreted by reference to a certain presupposed intuitive model M. Carnap avoids an appeal to such an intuitive model in the discussion of Language II itself, but he could not avoid it in the discussion of its syntax; for the conclusions belonging to its syntax would not be acceptable to Carnap*, though Carnap and Carnap* would, of course, always agree with respect to those conclusions which depend exclusively on formal considerations.[5]

Beth himself proved a major metamathematical result aimed at the pretense that the sense of mathematical concepts is somehow fully conveyed ("implicitly defined") by the axioms of the relevant extant theory. He explicated this view by proposing the principle:

> if F and F′ are implicitly defined by essentially the same axiom sets A and A′, then it must follow from A and A′ together that F = F′

and then proved

> V. (Beth) If F is implicitly defined by the theory T, then F is explicitly definable in T.

When F is explicitly definable in T, some expression in which F does not occur does exactly the same job as F there, so that F can be deleted without loss from the primitive terms of the theory. The significance of these results is a main area of contention between realists and antirealists. I can identify two other such areas subordinate to these, which I shall discuss in the next section.

The question most important to me is: Just what advantage is the mathematician supposed to reap from his purported ability to refer unambiguously (in an extreme *or* moderate sense) to "intended models"? This is in part the old question of what *access* the mathematician has to these independently existing objects. Does he know them by acquaintance or by description? And if the latter, is the description anchored in acquaintance with something, or are we to know those objects *solely* as "that which satisfies the axioms"? While only the most extreme Platonist would appeal to *Wesenschau,* it must be admitted that Husserl correctly identified a task which the mathematical realists have left woefully incomplete: to describe how these nonconcrete entities enter our experience.

IV has lately caused some activity which might provide a clue to mathematicians' purported access to abstract entities. Do mathematicians, faced with the independence of the strong axioms of set theory and hence the logical possibility of adding some of their rivals to the weak axioms, attempt to ferret out what is *true?* Gödel answered in the affirmative: "The mere psychological fact of the existence of an intuition . . . suffices to give meaning to the question of truth or falsity . . . ."[6] But my impression is that mathematicians are so busy exploring which extensions of the weak axioms are, to use Gödel's own terminology, "fruitful" and which "sterile," that the question of truth, or correspondence to a previous intended model, is not considered.

Of course, realists have their account of these phenomena too. Perhaps the situation should be compared to the rise of non-Euclidean geometries. Around 1800, everyone thought of actual space (physical space) as Euclidean, and they continued to do so for a while even after the development of other geometries. But then it became clear that there remained the substantive question of which geometry was the true one, the geometry of actual space. The situation now is similar. There are Cantorian and non-Cantorian set theories, but there is exactly one actual or correct milky way system, and at most one of the rival set theories can describe it correctly.

Perhaps I belabor the obvious if I point out that the question of which is the true geometry is a question that makes sense only if we identify physical correlates of geometric concepts, for example, if we stipulate that paths of light rays shall be geodesics. Is the realist ready to lay down a coordinative definition for ϵ? Or does he imagine that Riemann could have settled the problem by saying that he meant "straight line" in a straight line sense of *straight line?*

4. Subordinate Sound and Fury

Since the realist typically believes that at least the weak axioms of set theory are literally true, he (also typically) believes that there exist more than denumerably many entities. But the Löwenheim-Skolem theorem shows that no matter how he phrases his existence postulates, it will be consistent to say that each of his postulates is true, rightly understood, although there are still only denumerably many things in the world. (The point is simply that "denumerable" is a concept which varies depending on the domain of discourse; what is not denumerable in one domain may be denumerable in another.) So an antirealist, for example, a nominalist, may say that at the least there is no reason to believe that there are more than denumerably many entities in the world.

At exactly this juncture, the realist's gambit of placing the onus of proof on his opponents is most typically realized. I can prove the Löwenheim-Skolem theorem, he says smugly, but can you? Don't come back until you have developed a nominalistic mathematics powerful enough to prove such theorems. And don't try to appropriate by theft what I gained by honest toil.

But this victory-by-default gambit is simply beside the point. Recall Anselm's venerable strategy of beginning with "Even the fool sayeth in his heart. . . . " On the nominalist position there is no reason to believe that there are nondenumerably many things, since the nominalist denies the existence of abstract entities. On the conceptualist position there is no reason to believe that there are nondenumerably many things. (According to the intuitionist, Cantor's theorem shows only that the concept of the continuum is so rich that its sense is not exhausted by any single humanly followable recipe.) And on the realist position there is no reason to believe that there are nondenumerably many things, since the realist can show via Skolem's proof that his postulates can be satisfied without believing that. So even on the basis of his own position, the realist can find no justification for his belief.

Finally, there is no reason to believe so much. At this point it may be relevant also to mention the substitution interpretation of quantifiers. Not much is to be gained by it, since that still does not show that there is no reason to believe that there are more than finitely many things. But it is one of those subjects where those people who have found the realist's manner of posing the problems inescapable, then found themselves thereby hard pressed. Quine held that one could tell a person's ontic commitments by looking at his syntax and logic; I have found no more precise and tenable way of explaining Quine's criterion of ontic commitment. Arguments showing that Quine's own explications are either confused or inconsistent are familiar to everyone. Henkin (in 1953) and Sellars (in 1960) argued that it is possible so to use the substitution interpretation of quantifiers that one's syntax and logic give no clue to his ontic commitment.[7] Quine's retort that the substitution interpretation validates the Carnap rule '$Fx_1, \ldots, Fx_m, \ldots$ hence (x)Fx', where '$x_1, \ldots, x_k, \ldots$' are all the singular terms of the language, is not cogent. It was already forestalled in Henkin's 1953 paper by a point explicitly appreciated by Sellars: in the interpretation of the quantifiers in language L the set of substitution instances need not be defined by the set of singular terms of L itself. And in fact, one's syntax and logic are so far from betraying his ontic commitment that they do not even show whether he accepts the principle of bivalence.

Yet, as Henkin also pointed out, this is not a line of thought by which the nominalist can hope to reconstruct mathematics. If he tries, he will find himself trying, by implication, to provide the finitist proof of the consistency of arithmetic, which Gödel's results ruled out. Putnam suggests that to escape this dilemma the nominalist would be driven to consider *possible* singular terms and *possible* languages, an extremity that would be anathema to the nominalist (though not to a conceptualist).[8]

A second major realist tactic is to claim that the skeptic about abstract entities, having already thereby begun to topple from philosophical common sense, cannot help but fall headlong into skepticism *tout à fait.* His plunge is precipitated, it is argued, because by reason of analogy and because validity is a matter of form, skeptical arguments about abstract entities become skeptical arguments about physical objects. There are two examples of this line of argument.

When Beth comments on the impossibility in principle of developing categorical mathematical theories he says:

> This is at the very least strongly reminiscent of the situation in which we find ourselves with respect to physical objects. Even the most complete physical theory does not exhaust our intui-

> tive knowledge of physical objects. This gives us the conviction that physical objects are not mental entities, somehow created by the physical theory, but that they have an existence independent of any theory, or as we express it more briefly, they are "real".[9]

This is ingenious, but it think it assimilates the problem of not having uniquely identifying descriptions to that of not having complete descriptions (in each case, in principle). We would not be worried by the lack of complete theories if we had some other possible means of finding uniquely identifying descriptions of the supposed objects.

In a more obvious vein, Gödel suggests the argument:

> It seems to me that the assumption of such [mathematical] objects is quite as legitimate as the assumption of physical bodies and there is quite as much reason to believe in their existence. They are in the same sense necessary to obtain a satisfactory system of mathematics as physical bodies are necessary for a satisfactory theory of our sense perceptions and in both cases it is impossible to interpret the propositions one wants to assert about these entities as propositions about the "data", i.e. in the latter case the actually occurring sense perceptions.[10]

I wonder what "in the same sense" means here. If Gödel wants an account of mathematics that parallels the account of physics, he is clearly right, given his desiderata. If there is some necessity independent of such a prior choice of desiderata, he has not spelled that out. But let me put the question bluntly: Suppose one realist wishes to know what another uses the term 'the null set' to refer to. Well, how does he know to what the other refers by "the door" or "the Empire State Building"?

What is the force of such arguments anyway? Are they not simply another way to put the onus of proof on the opponents of realism? How cogent is it to tell antirealists to get themselves to a nunnery and develop an adequate epistemological account of how persons relate to physical nature before they dare to raise sceptical doubts about entities outside nature?

5. What the Realist Does Not Know

Epistemology has always been Platonism's Achilles' heel. The extreme Platonist combines a realist ontology with a theory of knowledge that implies our having direct acquaintance with abstract entities. We know their existence, he holds, not through abstraction, inference, conception, or confirmation of hypotheses, but as objects of intuition. This is by far the most comfortable epistemology for a realist to have; it suffers only from not having received a respectable elaboration for its own sake in several centuries.

As with most philosophical issues, the opposite extreme is presently espoused, with admirable daring and skill, by Putnam.[11] He proposes to combine a realist ontology with a purely empiricist epistemology. Mathematical entities, he holds, are epistemologically on a par with the theoretical entities of physics. Mathematical practice is, or reasonably can be, hypothetico-deductive, and its hypotheses may be confirmed by the canons of inductive reasoning. On his view, it is reasonable to accept Fermat's last theorem and Goldbach's conjecture, since we have been testing them exhaustively, and they have survived the testing process. This is one consequence of his view, and he draws it unblushingly. A second feature central to his account is that truth does not consist merely in correspondence to the facts, but has pragmatic aspects: the survival of one theory in the struggle of competing theories within the community of mathematicians (or physicists) is a mark of truth. This seems to leave open the possibility that more than one theory might correspond equally well to all the facts, and still exactly one of them be true. (Else, what difference is there?) This is a move one expects of a conceptualist rather than a realist, but the attempt to combine realist ontology with empiricist epistemology is in any case a heroic one.

Anyone who has to defend either such extreme epistemological doctrine is not in an enviable position, but the moderate realist who does not try to make his ontic commitments palatable through epistemological extremism is also in a difficult spot. He does not, for instance, know the abstract entities he purports to refer to by acquaintance. He can find no uniquely identifying descriptions for them, even in principle. Perhaps it looks as if he could for finite or even denumerable sets. The null set, for instance, is the set without members. But one tenable realist position is that sets are logical constructions from properties for which extensionality does not hold, and the theory of such properties is not uniquely determined by the theory of sets. But then if realist *X* holds it, realist *Y* can exhibit *X*'s properties as set-theoretic constructs, and *X*'s sets as constructs of constructs, in which case he says: X's null set is my thingamadiddle. So even if this realist knew his referents, he could not tell us what they are, even in principle. It is hard not to conclude that the realist does not know what he is talking about.

Of course, I only mean this in the purely literal sense that he does not know what he is referring to.

6. *The World of Id*

Could we possibly be living in the world of Id? Certainly realism is rampant; philosophers tend to believe in the existence of almost anything, and mathematicians talk as if they do. Would it have made any difference to the development and usefulness of mathematics if, in addition, there actually were no abstract entities? Is Id really possible? Anyone who says not ought to produce an ontological proof of the existence of the null set, at the very least. ("That than which no emptier can be conceived," anyone?)

Is mathematics possible in Id? Anyone who says not ought to produce a transcendental deduction of the existence of mathematical entities. But suppose someone did. Interpretations of the import of such deductions vary. One interpretation is that it would establish the existence of mathematical entities as a presupposition of the development of mathematics. In that case, would it not suffice to have the aspiring mathematician take an oath to keep presupposing all that? And would the oath, or his presuppositions, be any the less efficacious if this were Id? Would he ever notice anything wrong?

I am not arguing that there are no sets. First, it is philosophically as uninteresting whether there are sets as whether there are unicorns. As a philosopher I am only interested in whether our world is intelligible if we assume there are no sets, and whether it remains equally intelligible if we do not. Personally, I delight in the postulation of occult entities to explain everyday phenomena, I just don't delight in taking it seriously. As a philosopher, however, I look forward to the day when we shall be able to say, "Yes, Virginia, there is a null set," and go on to explain, as the *New York Sun* did of Santa Claus, that of course there isn't one, but still there really is, living in the hearts and minds of men—exactly what a conceptualist by temperament would hope.

5: R. M. MARTIN On Some Prepositional Relations

QUIS? QUID? UBI?
QUIBUS AUXILIIS? CUR?
QUOMODO? QUANDO?

As Russell noted already in *The Principles of Mathematics,* it is characteristic "of a relation of two terms that it proceeds, so to speak, *from* one *to* the other. This is what may be called the *sense* of the relation" (p. 95). The italicized prepositions here are of special interest, a salient feature of a dyadic relation being its sense. *From* and *to,* along with others, should perhaps themselves be regarded as primitive dyadic relations. If the logic of such relations were fully developed and clarified, no other relations need then be admitted either primitively or as values for variables. This at least is the thesis to be explored here.

According to the *O.E.D.*, always a good starting place in linguistic matters, a *preposition* is a part of speech serving "to mark the relation between two notional words, the latter of which is usually a substantive or pronoun." The confusion of use and mention here need not detain us–prepositions surely are not to be regarded just syntactically as marking relations between mere words. They seem rather to mark relations between or among nonlinguistic entities or things. Just what the "things" are here may include not only substantive individuals but also events, acts, states, processes, and the like. All such entities may be handled within the event logic developed elsewhere.[1]

According to another important entry in the *O.E.D.*, a relation in the most general sense is "that feature or attribute of things which is involved in considering them in comparison or contrast with each other; the particular way in which one thing is thought of in connection with another; any connection, correspondence, or association, which can be conceived as naturally existing between things." If prepositions are parts of speech construed as marking relations between things, what prepositions stand for then are the very relations (real or virtual) themselves. Such a relation is called here a *prepositional relation.* Let us reflect upon a few such relations as incorporated within suitable structures of English sentences.

Consider now

(1) 'John kisses Mary'.

Although the sentence contains no preposition, within event logic its structure in fact does. Let '⟨K⟩ *e*' express that *e* is a kissing act, '*e* By *j*' that j (John) is the *agent* of *e*, and '*e* Of m' that m (Mary) is the *recipient* or *patient* of *e.* Then one reading of (1) is

(1′) '(E*e*) (⟨K⟩ *e* · *e* By j · *e* Of m)',

that there is a kissing act by j of m.

There are other readings of (1) to the effect John is *now* kissing Mary or that John *frequently* kisses Mary. The first brings in time and the second the numerosity word 'frequent' as applied to virtual classes of acts. Let 'Now *e*' express that *e* is a present event.[2] The second reading of (1) becomes then

(1″) '(E*e*) (⟨K⟩ *e* · Now *e* · *e* By j · *e* Of m)',

and the third,

(1‴) 'Frequent $\{e \ni (\langle K \rangle e \cdot e \text{ By j} \cdot e \text{ of m})\}$',

to the effect that the numerosity of the virtual class of kissings of Mary by John is frequent. (In general, where '—*e*—' is a sentential form containing '*e*' as its only free variable, $\{e \ni —e—\}$ is the virtual natural class of all *e*'s such that —*e*—.)

The predicate '⟨K⟩' requires comment. Ordinarily one speaks of the *relation* of kissing, kissing being a "connection, correspondence, or association . . . conceived as naturally existing between things." The proposal of the present paper, however, is to handle all nonprepositional relations rather in terms of event-descriptive predicates or virtual classes of events, acts, states, or whatever. The only relations admitted primitively are thus the prepositional ones. Ultimately, of course, a complete list of all prepositional relations should be given, although this will not be attempted here.

A rather far-reaching change in the logic of relations ensues. Given any erstwhile nonprepositional relation R, '⟨R⟩ *e*' now expresses that *e* is an R-event, -state, -act, or whatever. In some cases one would speak of an act, as above. In others, of a state, as in loving. The structure, for example, of 'John loves Mary' is to the effect that there is a loving *state* on the part of John with respect to Mary, with additional clauses if needed, for 'John loves Mary now' or 'John often loves Mary', and so on. Every state or act is regarded as an event, in the broad sense of 'event' being used here.

Among nonprepositional relations, monadic ones may be included. A monadic relation is merely a virtual class, and these likewise may be handled by means of event-descriptive predicates. Where L is the virtual class of living beings, '⟨L⟩ *e*' may express that *e* is a living state or process.

(2) 'John is now living'

becomes then

(2′) '(E*e*) (⟨L⟩ *e* · Now *e* · *e* Of j)'.

Any one prepositional relation may be used in many different ways. In the structure for (1), 'Of' is used to express patiency, so to speak, whereas in (2′) it is used to express possession. The *O.E.D.* lists sixty-three distinct uses of 'of' in English, not all of them perhaps independent and some of them only in the context of a longer (usually prepositional) phrase. For the present, all the noncontextual uses of 'of' are to be accommodated by the prepositional relation Of. In a deeper study, however, it would no doubt be advisable to distinguish separate relations Of_1, Of_2, and so on, and ultimately to give meaning postulates concerning each. The Of here is then presumably a certain kind of relational disjunction of Of_1, Of_2, and so on. Similar remarks with regard to other prepositional relations to be considered are also to be presumed made.

Prepositions are often used merely as auxiliaries in relational words or phrases as, for example, in 'brother of' or 'greater than'.

(3) 'John is a brother of Mary'

becomes here

(3′) '(E*e*) (⟨Br⟩ *e* · *e* By j · *e* Of m)',

or perhaps

(3″) '(E*e*) (⟨Br⟩ *e* · *e* From j · *e* To m)',

to the effect that there is a brotherhood state going from John to Mary. And

(4) 'John is the one (and only) brother of Mary'

becomes

(4′) '(Ee) (⟨Br⟩ e · e From j · e To m · ~ (Ep) (Ee') (⟨Br⟩ e' · e' From p · e' To m · Per p · M p · ~ p = j)',

that is, there is some brotherhood state from John to Mary but no brotherhood state to Mary from any male person other than John.

The use of 'From' and 'To' here is especially useful in the case of dyadic acts or states indicating direction or sense, but 'By' and 'Of' could be used also. In a full working out of prepositional theory, it would be good to settle on one preposition to designate the relation between the act or state and the agent, another the relation to the patient, another to the direct object, another to the place, another to the time, and so on. Then the structures would be given exclusively in terms of these chosen few, and only these few need then be axiomatized via meaning postulates. The grammatical study of prepositions in English would then center around the formulation of rules for the transformation (or paraphrase, or translation) of sentences containing them into structures containing only one or more of these chosen few prepositions.[3]

'Jealous of' and 'gives' are usually taken to designate triadic relations.

(5) 'Edwina is jealous of Angelica's affection for Bertrand'

would normally be expressed simply as

(5′) 'Jeab'.

But according to the view given here, all relations are to be introduced in terms of corresponding relational states, acts, and the like. Thus the deep structure of (5) is not (5′) but rather something like

(5″) '(Ee) (Now e · ⟨J⟩ e · e From e · e To a · e Of b)'.

Then (5′) with a 'now' clause added may be taken as definiendum with (5″) as definiens. Similarly, of course, 'Br jm' may be taken as definiendum with (3′) or (3″) as definiens. It seems clear that any erstwhile relational name or constant can be introduced by definition in this way.

(6) 'Angelica gave the book to Henry'

becomes now

(6′) ‘(E*e*) (E*e*′) (Now *e*′ · *e* B *e*′ · ⟨G⟩ *e* · *e* By a · *e* Of (the book) · *e* to h)’.

(6) is logically equivalent to ‘The book was given by Angelica to Henry’, ‘The book was given to Henry by Angelica’, ‘Angelica gave (to) Henry the book’, ‘Henry was given the book by Angelica’, and to ‘Henry was given, by Angelica, the book’. The structures for all of these may be given merely by commuting various of the conjuncts of (6′). Thus, the passive forms of verbs come out very naturally, and therewith all manner of converses of relations. (Note that word order is regarded here as significant in linguistic structures. It is not strictly necessary to require this, but surely no harm results.)

The view expressed here concerning prepositions has some kinship with that of the linguist Charles Fillmore concerning grammatical cases. “I believe,” he writes, “that human languages are constrained in such a way that the relations between arguments and predicates fall into a small number of types. In particular, I believe that these role types can be identified with certain quite elementary judgments about the things [events?] that go on around us: judgments about who does something, who experiences something, who benefits from something, where [and when] something happens, what it is that changes, what it is that moves, where it starts out, and where it ends up. . . .”[4] The chosen few prepositional relations admitted as basic would have to be such as to enable us to specify the various role types to which reference is required for every event-descriptive predicate as occurring in all possible contexts. *De jure* this is simple enough, but vast galaxies of empirical data must be garnered to make such specifications adequate *de facto.*

Intensional relations are to be handled, as in *Events, Reference, and Logical Form,* by bringing in the Fregean *Art des Gegebenseins.* Here the suitable prepositional relation is *Under,* intentional contexts being those in which a given entity is taken under a mode of linguistic description. But ‘Under’ is readily definable in terms of reference. Let ‘*p* Ref *ae*’ express that the person *p* uses the linguistic expression *a* to refer to *e.* Then clearly ‘*e* Under_p *a*’ may abbreviate ‘*p* Ref *ae*’. An entity *e* is taken by *p* under *a* just where *p* Ref *ae.*

Consider 'intends' itself in a suitable context.

(7) 'Oedipus intended to marry Jocasta'

becomes something like

(7′) '(Ee) (Ee_1) (Ee_2) (e_1 B e_2 · Now e_2 · ⟨Intd⟩ e_1 · e_1 By o · e_1 To e · e Under$_o$ a · a = '{e' ∋ (⟨M⟩ e' · e' By o · e' Of j)}')',

to the effect that there is some e_1 before some present e_2, namely an intending by Oedipus to do some e under the linguistic description of its being a marrying by himself of Jocasta. 'Intd' here stands for the intensional relation of intending to do under a given description and is itself definable in terms of '⟨Intd⟩' as follows:

'p Intd$_q$ e,a' abbreviates '(Ee_1) (⟨Intd⟩ e_1 · e_1 By p · e_1 To e · e Under$_q$ a · Per p · q PredConOne a)'.

The e_1 here is an intending by person p to do e, where e is taken by q under the description a, q taking a as a one-place predicate constant.

Concerning the relation Intd, a *Principle of Intending under Paraphrastic Descriptions* such as the following presumably obtains:

(Ec) (Ed) (Ed') (p PredConOne a · p PredConOne b · p InCon c · p C dac · p C $d'bc$ · p Prphrs dd') ⊃ (q Intd$_p$ e,a ≡ q Intd$_p$ e,b).

Here 'p InCon c' expresses that p takes c as an individual constant, 'p C dac' that p takes d as the concatenate of a and c, and 'p Prphs dd'' that p takes d to be a paraphrase of d'.

Note the interesting distinction between 'p Intd$_q$ e,a' and 'p Intd$_p$ e, a'. The former expresses that p intends to do e under q's description of e, the latter under p's own description of it. Oedipus intended to marry Jocasta, the object of his intention being expressed in just this way either in his own vocabulary or in that of the speaker. Oedipus, however, presumably did not intend to marry his mother in his own vocabulary, so to speak, although he perhaps did in that of the speaker. The sense in which Oedipus did truly intend to marry his mother may be expressed by taking the object of Oedipus's intention under a description in the speaker's vocabulary. In other words, Oedipus did not intend to marry the person whom *he* describes as 'Oedipus's mother', but he did intend to marry a certain person whom the speaker is privileged to describe in this way if he wishes. To provide for this difference in the case of intensional words is a merit of the present notation.

If the foregoing considerations are sound, a very considerable simplification and a reduction in the logical theory of relations are achieved. It is presupposed that all relations are regarded as virtual and are thus in no way values for variables. But all primitive relations are now prepositional. In place of an erstwhile primitive nonprepositional relation R, an event-descriptive predicate '⟨R⟩' is introduced for the virtual class of all R-happenings, -events, or whatnot. Further, most, perhaps all, the primitive prepositional relations seem to be *dyadic* relations between events. (To be sure, some prepositional relations of higher degree seem to be definable. An example is perhaps 'between', which will be considered in a moment.) Thus all predicates are ultimately definable in terms of one-place event-descriptive predicates for virtual classes, a handful of two-place predicates for prepositional relations, and the resources of event logic. Within such a framework, the full elementary (first-order) logic of relations finds its place, of course, the point being that no specifically nonlogical relational primitives, other than the prepositional ones, are required.

Consider

(8) 'John now sits between Harry and Mary'

and its putative structure

(8′) $`(\mathrm{E}e)\,(\mathrm{E}e_1)\,(\mathrm{E}e_2)\,(\mathrm{E}e_3)\,(\mathrm{E}e_4)\,(\mathrm{E}e_5)\,(\mathrm{E}e_6)\,(\langle \mathrm{S}\rangle\, e \cdot \mathrm{Now}\, e$
$\cdot\, e\ \mathrm{By}\ \mathrm{j} \cdot \langle \mathrm{j},\mathrm{At},e_1\rangle\, e_4 \cdot \mathrm{Pl}\, e_1 \cdot e_1\ \mathrm{Bet}\ e_2,e_3 \cdot \langle \mathrm{h},\mathrm{At},e_2\rangle\, e_5$
$\cdot\, \mathrm{Pl}\, e_2 \cdot \langle \mathrm{m},\mathrm{At},e_3\rangle\, e_6 \cdot \mathrm{Pl}\, e_3 \cdot \mathrm{Now}\, e_4 \cdot \mathrm{Now}\, e_5 \cdot \mathrm{Now}\, e_6)'$,

to the effect that a present sitting state by John is such that the place occupied now by John is between the places now occupied by Harry and Mary. Here At is the relation of occupying a place, and Bet is the betweenness relation among events. Clearly, however, 'e' At e_1' may be regarded as short for '$(\mathrm{E}e_4)\,(\langle\mathrm{At}\rangle\, e_4 \cdot e_4\ \mathrm{By}\ e' \cdot e_4\ \mathrm{To}\ e_1 \cdot \mathrm{Pl}\, e_1)$'. And also perhaps '$e_1$ Bet e_2,e_3' may be regarded as short for '$(\mathrm{E}e)\,(\langle\mathrm{Bet}\rangle\, e \cdot e\ \mathrm{Of}\ e_1 \cdot e\ \mathrm{From}\ e_2 \cdot e\ \mathrm{To}\ e_3 \cdot \mathrm{Pl}\, e_1 \cdot \mathrm{Pl}\, e_2 \cdot \mathrm{Pl}\, e_3)$'. Place e_1 is between places e_2 and

e_3, provided there is a betweenness state e *of* (or possessed by) e_1 and holding *from* e_2 *to* e_3. If this definition is feasible, then 'Bet' need not be taken as a relational primitive, and similarly no doubt for Bet taken as a relation among entities other than places.

There is a difference to be noted between (8) and

(9) 'John is now sitting down between Harry and Mary'

or

'John now sits down between Harry and Mary'.

Sitting down is an act, whereas sitting is a state. In any case, '⟨S⟩', denoting states, should be distinguished from '⟨SD⟩', denoting acts of sitting down.

The question arises as to whether the logical primitives '=' for identity, 'P' for the relation of part to whole, and 'B' for the before-than relation, may also be introduced in terms of prepositional relations. Let '⟨=⟩ e' express for the moment that e is an identity state. Then '$e_1 = e_2$' is definable as '(Ee) (⟨=⟩ $e \cdot e$ From e_1 · e To e_2)'. And similarly for 'P' and 'B'. There seems to be no cogent reason why identity states, part-whole states, and before-than states should not be admitted along with brother-of states, loving states, sitting states, and the like. (In fact, the distinction between primitive logical and nonlogical predicates has always seemed rather specious anyhow.) On the other hand, there does seem to be a difference in extent of removal from natural language, in which nonlogical states and acts are commonly referred to, logical ones (so-called) rarely if ever. For the present the matter need not be decided, closely related as it is with the difficult question as to whether there is any fundamental sense in which logical and nonlogical constants differ.

The whole theory of virtual relations is of course preserved on this basis, relations being now available by definition.[5]

The unwary reader might think at this point that too much has been expunged, and that a full theory of (real, not merely virtual) relations is needed for the foundations of mathematics as incorporated within a type or set theory. The view here, on the contrary, is that the various areas of mathematics arise from the study of certain kinds of human *acts,* such as counting acts and acts of subjunctivizing, as discussed in *Events, Reference, and Logical*

Form. And anyhow the view seems now widespread among logicians, in the wake of Tarski, that type and set theory themselves have failed utterly to provide adequate foundations for mathematics—a view some of us have been urging all along. It is thus perhaps timely to explore other, more promising approaches. In any case no support can be given at this date for the theory of (real) relations on the grounds of its alleged need in type- or set-theoretic foundations for mathematics.

This brief sketch of a theory of linguistic structure needs expansion in several directions. A notation for reference and truth may be provided in part on the basis of 'Ref' above. A theory of pronouns and demonstratives, a theory of modal and epistemic words, a theory of intentionality, a theory of numerosity words, and a theory of mass nouns, to say nothing of adjectives and adverbs, may all be provided for here. These various topics have been treated to some extent in *Events, Reference, and Logical Form,* and nothing new in principle would emerge from viewing them anew in the light of the foregoing. This could easily be done with only minor changes.[6]

The exact study of linguistic structure is still in its infancy. It is by now surely clear, however, that the riches of first-order logic as extended in various ways can no longer be disregarded by the philosophical analyst or the professional linguist. Whatever its defects, the foregoing sketch of a system of linguistic structure is probably the simplest and most extensive that has yet been suggested. In any case, it is hoped that it will be useful as a basis for further study, extension, and improvement.

6: JOHN T. KEARNS

Sentences and Propositions

Propositions as Well as Sentences

1. Sentences Aren't Enough

In this paper I will examine the distinction between sentences and propositions and attempt to discover what the status or nature of propositions is. In discussing language, I will consider only those declarative sentences that can be used to make assertions, but of these I will *not* deal with sentences like 'I assert that the cat is on the mat', where a "performative" prefixes the content of the assertion. I focus rather on sentences like 'The cat is on the mat', not ruling out sentences that fail because they have a presupposition that is either not true or not satisfied. Throughout I shall avoid the issue of whether sentences containing nondenoting (empty) singular terms are false, or neither true nor false.

Many philosophers who discuss the problems and issues I shall consider here do not approach them in terms of the sentence–proposition distinction. Some prefer statements to propositions, and others incorporate such discussions into more general discussions of speech acts. A choice of one approach over another need not signal a substantive difference between the views presented. Given a topic or a problem, there are different ways to organize the study and discussion of it. One of these ways is not true while the others are false. One way can be more useful or helpful than another; and two ways may be equally satisfactory.

To justify postulating, or recognizing, propositions which are distinct from sentences, two sorts of examples are frequently used. The first sort of example is provided by two or more sentences, in the same or different languages, which have the same meaning. For example, the French sentence

(1) 'Il pleut'

means the same as the English

(2) 'It is raining'.

Some views have it that the proposition they express is what these sentences have in common. However, on such a view propositions would be unable to do all that is normally asked of them. For propositions not only get expressed, they are also made the bearers of truth and falsity. And neither truth or falsity can be associated with (1) and (2), or with their meaning, when these are considered in isolation. The sentences must be used before anything true or false has been said. And what a person says by using one of them on Monday is different from what he says with it on Tuesday. So propositions must be justified by different sentences used to make the same claim (used to say "the same thing"). Propositions must be more like what a sentence is used to say than they are like meanings.

The second sort of example often used to bring propositions to our attention is provided by sentences that can be used on different occasions to say different things. Sentence (2) used at different times or different places is used to make different claims. Although sentence (2) obviously does not mean the same as

(3) 'It rained in New York City on December 1, 1902',

sentence (2) could have been used to "say the same thing" as we can say by using (3) today. Armed with the sentence–proposition distinction, we can account both for different sentences being used to make the same claim and for one sentence being used to make different claims.

But just how do examples like these justify our recognizing propositions? When two sentences are used to say the same thing, there is clearly something common to the two situations, and this something *isn't* a sentence. We can simply *decide* to describe this case by saying that the two sentences (are used to) express the same proposition. And it is clear that what is common to the two situations is the same kind of feature that accounts for the difference when one sentence is used on different occasions to say different things. So the 'expresses a proposition' locution provides a convenient description for the two sorts of case.

The argument so far justifies using the word *proposition* for talking about sentences and their use, but it does not provide any information as to what "proposition" is a label for. Let us consider an attempt to explain propositions solely in terms of sentences—to reduce propositions to sentences. This attempt focusses on so-called "eternal" sentences, which are context-independent. Whenever different people use the same eternal sentence, they use it to say the same thing. If two different eternal sentences are used to say the same thing, we can try to account for this sameness in terms of the logical relations between the two sentences. And if these logical relations are syntactical relations, then sentences (and their structures) are sufficient to account for the sameness. But when used by someone, any context-dependent sentence will be equivalent to an eternal sentence. So the sentence–proposition distinction is reduced to the distinction between ordinary and eternal sentences. Unfortunately, this attempted reduction is unsuccessful. In order to recognize that some context-dependent sentence, when used, is equivalent to an eternal sentence, we must recognize that these sentences are (or can be) used to say the same thing. This is the very feature we want to understand, and we cannot use the fact that they do express the same proposition to explain what it is to express the same proposition. Nor will anything be gained by moving from eternal sentences in English (or whatever) to a language that contains only eternal sentences. If it happened that everyone spoke languages in which the sentences were all context-independent, then it might never occur to anyone to distinguish sentences from propositions. But the distinction would still be legitimate. Context-dependent sentences only serve to make it obvious.

2. Some Worries about Propositions

We must distinguish a sentence from what, on a given occasion, the sentence is used to say. But we don't know quite what it is that exists in addition to sentences. The word *proposition* has been chosen for marking the difference between sentences and what they are used to say, but we don't know what propositions are. The classical modern view of propositions is that they are timeless, nonlinguistic individuals which are the primary bearers of truth and falsity.[1] Propositions have no beginnings, they do not grow old or decay, and they do not change their truth values. If it is true that two plus three is five, then it is always (eternally) true. The same goes for George Washington being the first president of the United States. Propositions can't be tied to any one language, because sentences in different languages express the same proposition. Anyway, propositions outside of time can't depend on languages in time. Two plus three was five before there were any people saying anything.

Philosophers sometimes object to propositions, as classically understood, on the grounds that such propositions are abstract entities, and all abstract entities are objectionable. But the distinction between abstract and concrete entities is so vague that no one can be sure what he is doing when he rejects abstract entities. Consider concrete entities. What does it take to be a concrete entity? These entities can't be identified with objects of sensation, for no one writes off molecules as abstract entities. And we shouldn't identify concrete entities with physical objects, because if there were (nonphysical) minds (if there *are* minds), these would not be abstract. But if we can't say what counts as a concrete entity, then neither can we say what it is to be an abstract one. The abstract–concrete dichotomy may have some historically interesting or valid roots, but most of us seem to have lost sight of what these are. The expression "abstract entity" is now used as much to reveal the speaker's attitude as it is to characterize anything. So, I propose that the expressions "abstract entity" and "concrete entity" be dropped from the philosophic vocabulary.

Although I don't object to classical propositions for being abstract, I do object to their status because I am uneasy about timeless individuals and because I don't think propositions are nonlinguistic. I believe the basic conceptual scheme of ordinary language admits enduring individuals (continuants) which have properties and which are related to one another. These individuals have beginnings and ends. And the properties and relations of an individual can be different at different times. This is the basic ontological scheme presented, or represented, by our ordinary ways of talking about things. As basic, I think it should be accepted until it is proved to be untenable; and I don't believe it has ever been shown to be untenable. I am not going to argue for this position here. I am simply stating it to provide a point of reference for my discussion of propositions. I object to classical propositions because they don't fit into this scheme; these propositions are individuals which have properties and relations, but they aren't enduring individuals. My goal is to account for propositions in terms of the ontological scheme I have just described; to accommodate propositions within this scheme. To do this I will present a theory that allows us to distinguish propositions from sentences without admitting timeless individuals.

3. The Uses of Sentences

Classical propositions are suspect because of their extratemporal status. What is wanted is another way to accommodate the distinction between a sentence and what it is used to say. To begin with, we might note that for many of the things we say about propositions when we aren't talking about their ontological status, there are analogous things we can say about the ways sentences are used. Instead of saying that two sentences (are used to) express the same proposition, we can say that the two are used in the same way. And one sentence can be used in different ways.

If it were possible to account for the distinction between a sentence and what it is used to say by talking about sentences and their uses, then there would be no reason for recognizing the classical sort of propositions, for that distinction provides the only linguistic justification for them. But, in fact, the uses of sentences do the jobs for which propositions are wanted. Unfortunately, this can't be demonstrated by a short, neat argument. What is required is an adequate account of language in terms of sentences and their uses, one that resolves various problems about language and that avoids the difficulties associated with other theories.

The distinction between a sentence and the proposition it expresses is a distinction between a sentence and its use rather than a distinction between two individuals. When a speaker uses a sentence in a certain manner, the speaker and the sentence (token) are individuals, but the manner in which the speaker uses the sentence is a relation between the individuals rather than a third individual. For this reason it can be misleading to speak of a sentence expressing or being used to express a proposition; it would be more accurate to speak of a sentence and its *propositional use.* However, the original locution is often convenient. It must be understood as susceptible of more accurate elaboration.

Intentional Acts

4. Acts for a Purpose

To talk about sentences and their uses is to regard sentences as instruments that can be used to achieve various ends. This is an enlightening, and enlightened, way to regard sentences, but it lends itself to oversimplification and misunderstanding. In order to avoid this, I will devote considerable attention to intentional behavior. The intentional acts, or activities, I will consider are those done for a purpose; they are directed to some end.[2] I may sometimes talk as if all intentional acts were of this sort, but this is not quite true, for there are acts which are done *on* purpose but *for* no purpose. These do not include acts which are ends in themselves, for I am counting an act which is its own end as an act with a purpose. The only intentional acts I am not considering are relatively idle ones, and nothing of importance for this paper is lost by ignoring them.[3]

Any intentional act can be described, or regarded, in two ways. The first way takes account of the intentional character of the act. To describe an act in this way, one must use expressions suited to its intentional character. But an intentional act can also be considered as stripped of its intentional character. If my neighbor wants to go fishing and is digging for worms behind his garage, his behavior is intentional. My description of what he is doing has characterized his behavior as intentional. Even if I said that he is digging, but didn't say why, I would be characterizing his behavior as intentional. Digging is something that a person does for some purpose or other, and we take it for granted that he has a purpose. To strip my neighbor's activity of its intentional character I must describe what is going on behind the garage without placing that phenomenon in an intentional context. To do this I might describe the position of my neighbor's body, the physical relation of his body to the shovel, and the motions that are taking place. If I knew enough, I could indicate the physiological states and events involved. But my description cannot mention, or presuppose, that my neighbor is acting for an end, or even that what is going on is the kind of thing that generally constitutes intentional behavior. Normally, to say that a person is acting in some way or doing something is sufficient to locate his behavior in an intentional framework.

Let us call the two sorts of descriptions of an intentional act intentional descriptions and brute fact descriptions. It is possible that a single description might combine elements of both sorts of descriptions, but the distinction between intentional and brute fact descriptions is clear enough. For any act which gets an intentional description we can also frame a brute fact description. The two descriptions describe the same state of affairs, but they are not equivalent. For most purposes of daily life, intentional descriptions are more adequate than brute fact descriptions. When we ask what is going on in some situation where people are involved, we normally don't want an explanation in terms of motions and muscles.[4] Intentional explanations or descriptions are needed for a satisfactory account of language, for the use of language by human beings is intentional behavior.

Intentional acts can be more and less difficult, more and less complicated, more and less time-consuming. But if what someone is doing is too complicated or too time-consuming, we tend to speak of his activity rather than his act. When a dentist drills my tooth, it is his activity which causes me pain. Similarly, if Jones is building a cottage in his spare time, his building is activity rather than act. A certain activity might include many distinct acts, although acts can also include component acts. And we are more likely to speak of activity rather than of an act when we discuss a joint enterprise involving the participation of several people. But we do sometimes talk of joint acts. There does not seem to be a sharp dividing line between intentional acts and intentional activities; both acts and activities have the same intentional features. So I will regard acts and activities as roughly the same sort of things.

Let us restrict our attention to acts or activities performed by a single person. When someone acts intentionally, he acts for some purpose. Standard forms of expression for describing intentional acts, might be

(4) *X* does *Y* in order to obtain *Z*,

(5) *X* does *Y* in order to accomplish *Z*.

With a little forcing, we could probably make do with either (4) or (5), but each form fits certain acts better than does the other. Form (4) works best when it is having something after the action that matters most. Form (5) is appropriate when it is the doing (or being done) that is most important. In (4) and (5), *doing Y* is the act, and the purpose for doing it is *to obtain Z*, or *to accomplish Z*. The purpose for performing an act is not the same as the end or goal of the act. The act has its purpose while it is being performed, and it has a purpose whether or not it is successful. We can say that an act achieved its purpose or failed to achieve it, but the purpose isn't something one tries to get. The purpose for doing something is *to reach* the goal. The purpose for doing *Y* is *to obtain Z;* the goal is either plain *Z* or *to have obtained Z.*

There are two kinds of acts and activities that have some contrasting characteristics. (This is not to say that there are only two kinds.) If the difference between these two kinds of acts or activities is overlooked, characteristics proper to one kind may be attributed to all intentional acts. Acts and activities of the first kind are *cumulative.* They go on for a period of time, and at different moments during that period we would normally say that the agent is performing the act. My activity of walking to work is cumulative because I am walking to work during my whole trip. Building a house is cumulative. The term "cumulative" suggests activities that accumulate. These are activities that involve the doing of a certain number of things—the performing of a certain "amount" of activity. One keeps doing more and more until he is finally finished. But I also intend for the label "cumulative activities" to apply to intentional activities that don't accumulate. If someone walks in the country to get fresh air, his activity is cumulative. But there is no amount of walking he must accomplish; whenever he stops he will have walked to get fresh air. His activity is cumulative because it takes time, and all during that time he is performing the activity.

Some cumulative activities are interrrupted. Jones can be building a house on weekends and working at his regular job during the week. There are some situations in which, during the week, we would not say that Jones is building a house. These are situations where we are concerned with what Jones is doing at the time we are talking. In other situations it is proper, even during the week, to say that Jones is building a house. (We are concerned with Jones' long-term projects.) These two possible locutions (during the week) do not pose a problem for the present account of cumulative acts and activities. We understand what is going on (with respect to Jones) perfectly well. It is the speaker's purpose for communicating that determines which locution is appropriate.

The second kind of intentional act I want to call attention to is the act that is the completion, especially the successful completion, of some cumulative activity. For most cumulative activities there are no special expressions to label their completion. But we can distinguish performing an activity from completing it. However, there are some completing acts for which we have names.

Winning the race is the successful completion of running the race. Reaching London is the successful completion of travelling to London. And finding what one has lost successfully completes looking for what he lost. (These three examples are taken from Ryle's *The Concept of Mind.*) These completing acts are events, but they are events which are counted as intentional acts, for we consider a person to be *doing* something when he wins a race, reaches London, or finds what he was looking for. Completing acts are located in time, but they are not "spread out" in time like cumulative activities. A completing act need not be instantaneous, but we wouldn't normally speak of a person as performing the same completing act at recognizably different times.

The distinction between cumulative acts or activities and completing acts is important in considering what counts as the competent performance of an act and what it takes for an act to be successful. For most cumulative acts and activities, a reasonably competent performance is required before what someone is doing counts as an intentional act or activity of a certain kind. For to say that someone is doing *Y* (where doing *Y* is an intentional act) is to say that he's doing it right, or approximately right. This is so whether he's speaking French, building a house, or tying his shoe. There is often disagreement about the level of competence required and standards of competence, as in the case of artistic activity. But competence is a requirement for someone's performing a cumulative act or activity. With completing acts we don't "require" competence, for the performance of a completing act presupposes a reasonably competent performance of the activity that precedes it.

The competent performance of a cumulative act or activity is distinct from its successful conclusion. This is so even for those cumulative acts or activities whose competent performance coincides with their successful conclusion, as happens with some short-lived acts. If having the lights go on successfully concludes flipping the light switch, then this successful conclusion coincides with the intentional act of flipping the switch. There are relatively few cumulative acts or activities for which a competent performance completely guarantees a successful conclusion. But there are many cumulative acts and activities for which a competent performance is *normally* sufficient to insure success, as flipping the light switch is normally sufficient to get the lights turned on. And the competent performance of the activity of building a house is usually enough to get a house built. With completing acts we can't distinguish between the performance of the act and its successful conclusion. For the completing act completes or concludes a preceding activity. The preceding activity may be successfully concluded as in the case when a person finishes his running by winning the race. Or it may be unsuccessfully concluded, as with the runner who loses. But the completing act isn't itself completed or concluded. Of course, the completing act may have a purpose of its own, as in the case where Smith won the race in order to bring honor to his native land. This purpose may or may not be achieved, but even if it is, we would not consider this achieving to be the successful completion of the winning. Winning *is* a completion, it doesn't *have* one.

5. Complex Acts

We have seen that cumulative activities may not be concluded successfully, even when they are performed competently. Flipping the switch may not get the light turned on, and constructing competently may not get the house built. Language often provides a distinct label for cumulative activities brought to a successful conclusion where the label indicates that the act or activity was successful. We can say that Brown turned on the lights (by flipping the switch) if we want to make it known that his flipping succeeded. When language provides some label for an act or activity and another for that act or activity brought to a successful conclusion, it sometimes appears there are two acts or activities rather than one. For Brown can flip the switch in order to turn on the lights without getting them turned on; his act of flipping is the same no matter what its outcome. So flipping the switch is one act, and turning on the lights another. This suggestion that there are two acts rather than one is a suggestion I want to accept. And I will say that the first act is a part of the second. In order to understand how it is that one act or activity can contain others, we must now consider complex acts and activities.

There are many acts and activities which we ordinarily regard as complex, as containing other acts and activities. When Green carves the roast, his carving includes the act of putting the fork in the roast and various acts of slicing. And when a football player kicks off, he locates himself a certain distance from the ball and takes several steps before striking the ball with his foot. Acts and activities contained in a complex activity are not contained like bundles in a sack or elements in a set. The complex act has some structure; one act must precede another because it prepares the way for that other, or an act may wait on the completion of several other acts. What makes the several components of an activity to be parts of one activity rather than several ununited activities is that they aim at the same conclusion. Suppose I want to fasten two boards together with nails. I hold the boards in my left hand and hammer with my right. I hold the boards in order to fasten them together, and I hammer nails for the same reason. These two acts are joined by their common purpose, and this common purpose "defines" the act of fastening the boards together which contains these two acts. In a similar way, the different constructive activities that are part of building a house are united by being directed to the completed house.

It is not difficult to recognize a complex activity that has several distinct components. The complex acts that are most troublesome are those that contain a single act component. Consider this example:[5] O'Brien wants to blow up a bridge illegally. He plants explosives, sets a timing device to detonate the explosives, and leaves. There are two acts I want to consider, his planting the explosives and his blowing up the bridge. These are both acts done for a purpose, but their purposes are not quite the same. O'Brien planted the explosives in order to blow up the bridge. And, if his planting succeeds, his blowing up the bridge has the purpose to further the revolution (or to get revenge, or whatever). All the purposes for blowing up the bridge will also be purposes for planting the explosives, but the planting has one purpose that is not a purpose for the blowing up. The planting and the blowing up are clearly different acts. Not only do they have distinct purposes, but one can be performed without the other. O'Brien has planted the explosives, no matter what happens after he leaves. But if the explosives are discovered and disarmed, or if the timing mechanism or the detonator fails, then he hasn't blown up the bridge. On the other hand, if everything goes right, from O'Brien's point of view, and the explosion wrecks the bridge, then O'Brien has performed the act of blowing it up. The explosion that destroys the bridge successfully concludes O'Brien's planting of the explosives. The blowing up the bridge is the planting brought to a successful conclusion, just as turning on the lights is flipping the switch brought to a successful conclusion. But the time lapse between the planting and the explosion makes it obvious that the planting and the blowing up are different acts. And the purpose of the first act (the purpose is to blow up the bridge) defines the second act just as the common purpose of several acts defines the act that contains them. The complex act with a single component is simply a limiting case of complex acts. In all these cases the components are distinguished from the complex act by having a purpose which they don't share with the complex, and this purpose defines the complex act.

When someone does *Y* for some purpose and succeeds, we must recognize two acts: the competent doing of *Y* and the successful realization of the agent's purpose by doing *Y*. It may be that acts other than *Y* are also needed to achieve his purpose, so that doing *Y* isn't all that the agent must do; or it may happen that doing *Y* is the only contribution the agent needs to make. In either case, the successful realization of the agent's purpose is a complex act which contains doing *Y* as a component. Some cumulative acts or activities are successfuly completed by events which are not acts, as O'Brien's planting is completed by the explosion. And some are successfully completed by events which are acts, as running in the race is successfully completed by winning it. (But some successfully concluded acts may not be completed by any kind of event.) For those which are completed by non-act events, we commonly have distinct labels for the component act and the component act brought to a successful conclusion, as flipping the switch and turning on the lights. For these acts, the complex act includes both the component act and the completing event. When cumulative activities are completed by acts, there is frequently no name for the cumulative act brought to a successful conclusion. We don't seem to have, for example, a special label for successfully running in a race. This is probably because normal conversational demands can be met by talking of running and of winning. But the successful running must be recognized as a distinct activity which includes as components both the running and the winning. The purpose for successful running is also the purpose for winning; in fact, the purpose for winning is derivative from the purpose for running successfully. This explains why achieving the purpose for winning a race does not count as the completion of the winning. Winning has no purpose proper to itself; it is the successful running which may be further completed by achieving its purpose.

The distinction between completing and cumulative acts, as we have it so far, does not fit acts like blowing up the bridge very well. That act occupies a temporal stretch—from the time O'Brien began planting the explosives until the time of the explosion. But we wouldn't say that O'Brien is blowing up the bridge when he is planting the explosives, and we wouldn't say that he is blowing it up when he is sitting at home, waiting for the explosion. Part of the reason we don't say beforehand that O'Brien is blowing it up—in contrast with our describing Jones as building a house or Brown as writing a book before we know they will be successful—is probably due to the chancy nature of O'Brien's enterprise. A competent planting doesn't insure a bridge-destroying explosion, so we don't label O'Brien's act until the bridge is destroyed. But there is another reason why we will say ahead of time that someone is building a house but not that he is blowing up a bridge. Building a house is a complex activity that contains many components. The only thing that holds these components together is their common purpose, to build a house. The various constructive acts and activities don't constitute some single activity which has to build a house as its purpose. The complex activity of building a house is defined by the common purpose of its components. We say that Jones is building a house even before he is done, because we have no other convenient expression for labelling what he is doing, unless we say that he is *trying* to build a house. This is not the case with O'Brien's acts. Blowing up the bridge has a single act component, though the planting undoubtedly has components of its own. Since we can refer to his planting the explosives, we have no real need to describe him as blowing up the bridge before the explosion. Blowing up the bridge does not satisfy the earlier description of cumulative acts, but it clearly "belongs with" the cumulative acts described above. So I will extend the range of application of "cumulative act (or activity)" to include all intentional acts and activities which take some time and for which we would normally be willing to recognize different temporal parts. These are the most basic sort of intentional acts, for these are the primary "bearers" of purposes—as opposed to completing acts, whose purposes are derivative.

6. *Purposes*

To understand complex acts properly, we must devote more attention to the purposes for which acts are performed. Not only do purposes make intentional acts intentional, but they also provide the structure for complex acts. For the purpose of a component act distinguishes it from a complex act to which it belongs and defines that complex act. Someone who pounds nails with a hammer in order to fasten some boards together performs two acts: pounding nails and fastening the boards together. These two acts coincide temporally, but the first has a purpose that the second does not. This purpose, to fasten the boards together, furnishes a label for the second act. It is because the hammering has a purpose of its own that it can be performed when the fastening is not. For the hammering must be performed successfully for the boards to be fastened. If the nails are driven at an angle, then the hammering can be performed without the boards being fastened. (In this case, the pounding is a part of trying unsuccessfully to fasten the boards together.)

Most intentional acts have more than one purpose. If one act is part of another, the purpose of the complex act will also be a purpose of the component. And if the complex act is part of a still larger complex act or activity, the purpose of the larger complex act will be another purpose of the first component. Thus we are led to distinguish more and less remote purposes of an act. The least remote purpose of an act is its immediate purpose. It is its immediate purpose which enables us to distinguish an act from the complex acts to which it belongs. For the immediate purpose is the one purpose which it does not share with some complex act which contains it. Most acts are parts of other activities, and these are parts of still other activities. If, finally, we reach some complex activity that is part of no other, then its purpose can be considered an ultimate purpose. (But it seems that an activity must be self-sufficient, an end in itself, for its purpose to be ultimate.)

We have noted that the purpose of a component act "defines" the complex act to which it belongs. What this means is that we can, usually, transform the expression for the purpose of the component into a label for the complex to which it belongs. If someone flips the switch in order to turn on the lights, we can take the purpose expression "to turn on the lights" and transform it into the act expression "turning on the lights." And if we hammer nails in order to fasten two boards together, the label for our purpose can be transformed into the act (or activity) expression "fastening two boards together."[6] (But there may also be expressions labeling complex acts or activities not obtained from the purpose expressions for component acts; we can usually refer to a single thing in many different ways.) When a purpose expression is transformed into a complex act or activity expression, it labels the component act or acts brought to a successful conclusion, as O'Brien's blowing up the bridge is his planting explosives brought to a successful conclusion. The purpose of its components defines the complex act because the complex can be labeled by the purpose of its components, and because the complex depends on its components. The components are not performed in order to be parts of the complex. The purpose of the components is prior to the complex act, both logically and, usually, temporally.

7. *Instrumental Acts*

In performing intentional acts we frequently use instruments. When person *X* does *Y* in order to obtain *Z,* he may use instrument *A* in doing *Y.* Instrumental acts are a very important kind of intentional acts, for they are the ones we must understand if we are to understand the use of language. There are many different sorts of instrument and different ways to use them. Some instruments are used on something, or applied to it. As a hammer is used on or applied to a nail, and a pencil is used on a piece of paper. But other instruments are not applied to anything. As the tightrope walker who uses a pole to balance is not using the pole on something.

If we wish to adequately characterize an instrumental act, we must indicate what instrument is used, and what, if anything, it is used on. When doing *Y* is an instrumental act, and *A* is the instrument, we can replace 'doing *Y*' by 'using *A*,' so that we could write '*X* uses *A* (on *B*) in order to obtain *Z*' as our standard form for sentences. The parentheses around 'on *B*' indicate that instruments aren't always used on something. However, this standard form isn't adequate. To characterize an instrumental act completely, we must indicate more than the instrument used and the object it is used on. We need also to indicate the manner in which the instrument is used. For many instruments can be used in different ways. A claw hammer, for example, is designed for use in pounding nails and in pulling them up. And the hammer is used one way when it is used to pound and a different way when it is used to pull up.

To characterize an instrumental act completely, we must indicate the instrument(s) used, the object on which it is used, if any, and the manner in which the instrument is used. But we need to be more specific about what manner is intended here, for many different aspects of the performance of an act are part of the manner or manners in which it is performed. If we say that something was done well or poorly, we are evaluating a performance, but we are also characterizing the manner in which the act was performed. We do so as well when we say that something was done rapidly or slowly, easily or with difficulty. But these manners are clearly distinct from the manner of using an instrument that determines an act to be one kind of intentional act rather than another. To locate such manners as the manner of using a hammer to pound nails, I will speak of the "instrumental manner of use."

The amended standard form for describing instrumental intentional acts is thus 'X uses A φ-ly (on B) in order to obtain or accomplish Z.' (We could also build the 'on B' into φ, as in 'Jones uses the hammer nailpoundingly', which would yield the simpler 'X uses A φ-ly in order to obtain or accomplish Z.') This form provides a relatively complete analysis of instrumental acts, even though sentences of this form are frequently not very good English. English doesn't often provide us with convenient adverbs for indicating the (instrumental) manner in which an instrument is used. If we wish to distinguish the use of hammers to pound nails from the use involved in pulling them up, we can introduce a clumsy expression and speak of using the hammer *poundingly* on nails (or nail-poundingly). But it is possible to characterize instrumental manners of use without inventing barbarous adverbs. To do this, we can label the instrumental act without referring to the instrument used, and then convert this label into a phrase to indicate the instrumental manner of use. If doing Y is an instrumental act or activity and A is the instrument, we can say that X uses A in doing Y, or that X uses A to do Y. So now, instead of saying that Jones used the hammer nail-poundingly, we can say that he used the hammer in pounding nails, or that he used it to pound nails. Of the two locutions, I prefer the 'to do Y' locution and will use it as an alternative to the adverbial one.

The two standard forms for characterizing instrumental acts are 'X uses A φ-ly (on B) in order to obtain Z', and 'X uses A to do Y

in order to obtain *Z*.' The second form gives more natural English sentences than the first, as: "The acrobat used the pole to balance in order to walk across the tightrope without falling." But this form also lends itself to misunderstanding, for if we say that someone used *A* to do *Y*, this may suggest that he used *A* in order to do *Y*. In English it is common to have sentences like '*X* does *Y* to obtain *Z*' as a kind of "abbreviation" for '*X* does *Y* in order to obtain *Z*'. The sentence

(6) Joe clapped his hands to get Mary's attention

does "abbreviate"

(7) Joe clapped his hands in order to get Mary's attention.

In (6), "to get Mary's attention" does *not* indicate the manner in which Joe clapped his hands. But if we say that someone used a hammer to pound nails, this is *not* short for saying that he used it in order to pound nails. The expression "to pound nails" does serve to characterize the manner in which the hammer is used; it tells what he did with the hammer, not why he used the hammer. To determine whether an infinitive phrase indicates manner or purpose, we must decide whether or not using *A* to do *Y* is the same act as doing *Y*.

The purpose-manner distinction is further obscured by the fact that the very same 'to do *Y*' expression may be used on one occasion to indicate purpose and on another occasion to indicate manner of use. If doing *Y* is a complex act, and doing *M* is one of the principal act components, then we can speak of doing *M* in order to do *Y*. But doing *Y* may be an instrumental act, so that *A* is used to do *Y*. Suppose a person wishes to have a straight line connecting points *P* and *Q* on a piece of paper, and he decides to use a straightedge. His use of the straightedge involves two acts: holding the straightedge firmly in place with his left hand, and holding the pencil against the ruler as he draws. Now he holds the ruler in order to draw a straight line between *P* and *Q*. And he holds the pencil against the ruler as he draws in order to draw a straight line from *P* to *Q*. Using the straightedge *is* drawing a straight line. The expression "to draw a straight line between *P* and *Q*" indicates the manner in which the straightedge is used, although it labels the purpose for performing the component acts. The *purpose* for using the straightedge is to obtain a straight line between *P* and *Q*. The drawing is successfully completed by the existence of a straight line which connects the two points.

Applying the Analysis to Language

8. Propositional Acts

In an earlier section, the distinction between a sentence and the proposition it expresses was alleged to be reducible to that between a sentence and its propositional use. We must now see whether the lengthy discussion of intentional acts has clarified or supported this claim. To begin with, it is clear that certain expressions (tokens) are instruments which can be used in diverse manners to accomplish various purposes. Unlike most instruments, expressions don't exist prior to their use by a speaker or writer. The person who uses language to communicate must produce the expressions that he uses. However, normally there are not distinct intentional acts, one for producing and the other for using. The person produces and uses the expresson as one intentional act.

We shall now consider the acts performed by using sentences propositionally. Given the preceding discussion and theory, it might seem that sentences are instruments, and that when a person uses one propositionally he has performed what might be called a propositional act. But the situation is not quite so simple. There are two sorts of complicating factors. The first involve the components of a sentence and the components of the act of using a sentence propositionally. For a sentence is a complex instrument, and the act of using a sentence is a complex act. To fully understand such acts, we need a careful account of the acts performed by using the components of a sentence, and of the way these are unified into a complex act. The second set of complicating factors involves the immediate purpose of the act performed by using a sentence propositionally. Is this purpose something like to assert, to propose, or to suggest, so that one uses a sentence propositionally in order to perform what J. L. Austin called an illocutionary act? Or is there some intermediate purpose for using a sentence? This paper is not the place to solve these problems. Instead I will call attention to four alternative accounts, which exhaust the possibilities within the framework of the theory I have developed and show that each of them yields an acceptable view of the propositional use of a sentence—so far as ontological matters are concerned.

The first view is similar to the view John Searle has defended in arguing against locutionary acts (in "Austin on Locutionary and Illocutionary Acts," *The Philosophical Review* 77 (1968)). On this view, the components of a sentence are used to perform various acts—of referring, say, or predication. But these acts are combined into an illocutionary act like asserting. (The immediate purpose of the component acts defines an illocutionary act.) The second view is closer to Austin's original view. It recognizes an intentional act—a *propositional act*—performed by using a complete sentence. The propositional act is the act of using a sentence propositionally. The component acts are performed in order to use the sentence propositionally, and the propositional act is performed in order to assert something, or to suggest something, etc. (The propositional act is like a locutionary act, and the asserting, suggesting, etc., will be illocutionary acts.) On this second view, a sentence is used in a certain instrumental manner in order to perform an illocutionary act. On the first view, Searle's, we might also speak of propositional acts and propositional use of a sentence. But the significance of these locutions will be different. On the first view, talking of a propositional act is a way of talking about an illocutionary act in abstraction from its illocutionary force. A propositional act *is* an illocutionary act, but we talk abstractly about propositional acts so that we can emphasize what might be called the content of the act. Similarly, on the first view we can talk about the propositional use of a sentence as a shorthand way of talking about the various instrumental uses of the various components of the sentence.

The third and fourth views are derived from Frege. Both hold that a sentence is used to refer, or to do something like referring, to some such thing as a situation, fact, or truth value. Let us call such referring sentential referring. On both views, the immediate purpose for referring sententially would be to assert or to suggest, etc. The third view holds that the components of a sentence are used to perform acts whose immediate purpose is to refer sententially. On this view, the sentence is not a single instrument for performing a propositional act, i.e., the intentional act which consists in using the sentence in a certain instrumental manner. The sentence is rather a set of instruments plus some devices which indicate how the component acts are combined to yield the complex act of sentential referring. On this view, to speak of a propositional act would be an abstract way of talking about an act of sentential referring. On the third view, like the first, we speak of the propositional use of a sentence to abbreviate talk about the uses of its components. The fourth view is closest to Frege's actual view. It is a combination of the second and third views. On this fourth view, the immediate purpose for the component acts is to use the sentence propositionally. The immediate purpose for this propositional act is to refer sententially, and the immediate purpose for referring sententially defines an illocutionary act.

Although this paper is not the place to decide between these four views, we can see that all four allow us to speak of a propositional act which is the act of using a sentence propositionally. On all four views, a sentence is an instrument and/or a set of instruments. Using a sentence to express a proposition is using that sentence in a certain instrumental manner, or using its components in various instrumental manners. If we say the sentence "Snow is white" is used to express the proposition that snow is white, the phrase "to express the proposition that snow is white" indicates the manner in which the sentence is used. But it might be helpful if we drop the apparent commitment to an expressed individual and say instead that the sentence is used to "present" that snow is white.

We should note that it is not only the producer of a sentence who uses it. Although it is at odds with ordinary language to regard the person who listens or reads as acting, such a person is surely *doing* something. The spoken or written sentence must be "appropriated," it must be used by the receiver, if the producer is to succeed in communicating with that receiver. For the communication to be successful, the receiver must use the sentence or its components in roughly the same instrumental manner as the producer. But the receiver will not have a purpose that defines an illocutionary act. His corresponding purpose might be something like to learn what is the case, or to learn the speaker's view, or simply to be polite.

9. Ontological Matters

It might be objected to what I have done so far that I have not eliminated the undesirable features of classical propositions, for propositional uses, and all relations, are abstract entities. My goal should be to dispense with such things entirely. However, while it is true that propositional uses are abstract entities, whatever *they* are, this is no reason for rejecting them. Properties and relations are unobjectionable components of the world. They only make trouble when they are confused with individuals. For the difference between individuals on the one hand and properties and relations on the other is categorial. If one regards properties and relations as individuals, he is forced to introduce a new relation, sometimes masquerading under the alias "tie," and call it "exemplification" or some such. If he allows that this is a relation, then the process can be continued indefinitely.

Propositional uses aren't individuals, so they aren't "pieces" of language like sentences (sentence tokens). But they are linguistic entities, since the propositional use of a sentence is conventional. It is determined by the linguistic rules and conventions accepted and followed by members of a linguistic community. A language consists, among other things, of expressions and instrumental manners of use that are "assigned" to expressions. Given a sentence, there is a use or range of uses for that sentence. The sentence must be used in an appropriate manner if its use is to be correct and significant, but there is some flexibility on this score, as is shown by metaphors. And it is always possible to extend or enrich the language. Sentences in different languages can be used in the same instrumental manner. These manners, however, remain linguistic, for the instrumental manner of use is the manner of using some *expression.*

The primary reason why classical propositions have been regarded as nonlinguistic is their timeless character. So far we have not accounted for that character. To do that, I must find a "real" feature of propositional uses that is plausibly misconstrued as timelessness. To this end, it is necessary to consider in what sense there "is" a certain propositional use when no sentence is being used that way, or when no sentence is ever used that way. What sense does it make to say that some things which might be said will never be said? The answer to this is that an instrumental manner of use is a linguistic possibility; it is a possibility whether or not it is ever made actual, whether or not it is ever exemplified.

This sort of possibility is one of the basic kinds—perhaps it is *the* basic kind—recognized by ordinary language, and it is fundamental to a continuant ontology. But it is not much discussed by philosophers nor is it explicated or illuminated by the customary systems of modal logic. In this basic sense of "possible," something is possible *for* an existing individual or several existing individuals. As, for example, it is possible for John and Mary to have a baby: having a baby is possible *for* them. (More precisely, it is possible for them to cooperate in Mary's having a baby.) Propositional possibilities, e.g., the possibility that such and such, and possible individuals or states of affairs are derivative from this basic sense of possibility. It is possible *that* John and Mary will have a baby *because* having a baby is possible *for* them. Some things are possible for many different individuals or groups of individuals, as having a baby is possible for many couples, and not just for John and Mary. And some possibilities are "available" at many different times or for long periods of time. But nothing is possible for an individual that is defunct. And if they live long enough, there will come a time when having a baby is no longer possible for John and Mary. With this sort of possibility, something is possible for one or more existing individuals at some time.[7]

It is primarily properties and relations which are possible, in this basic sense, for individuals. For there are two ways that there can *be* a property or relation. There are two sorts of existence, two senses of "exist," which are enjoyed by properties and relations. The simplest sort is that which a property or relation has when it is exemplified by the appropriate number of individuals. The second way that there can be a property or relation is as a possibility for an existing individual or individuals. Properties possible for an individual can be more and less remote. Property φ may be possible for me now in such a way that by acting in an appropriate manner I can exemplify φ one minute from now. Property ψ may also be possible for me, but to exemplify ψ I must first exemplify φ, and only two days later will I exemplify ψ. It is possible for me both to learn Russian and to speak Russian, but I must learn it before I speak it.

A propositional use is a relation. it can either be exemplified or be available as an unexemplified possibility. The individuals needed to exemplify these relations are people and sentences. The sentences don't always exist beforehand, for in talking the speaker must produce the sentence so that the propositional use can be exemplified. So the propositional use is a slightly remote possibility for a talker; it depends on his actualizing some other possibilities. But the written or printed sentences in books, newspapers, letters, etc. are existing individuals waiting to be used. Propositional uses are possible for those sentences and the people who can read them. Some propositional uses are available to members of a linguistic community at a certain time. Given that community with its linguistic rules and conventions, there are many things which can be said. But there are also more remote linguistic possibilities. In order to say some things we may have to introuce new expressions and new manners of use. Although these things can't be said with the linguistic resources now available, they are genuine linguistic possibilities, for a natural language is never fixed and final. It is even conceivable that a whole new kind of language might need to be developed before some things could be said. But it is still possible to say these things, although the possibilities are remote. This doesn't mean that they are unlikely, but only that a lot of things must be done before they can be realized.

A sentence is true or false with respect to a propositional use. But we also say that what someone has said is true or false. The sentence isn't *what* he said; the sentence is the instrument he used to say it. When we say that what he said is true, we aren't characterizing his act. For, normally, when we say that what he said is true we are talking about what could be said again. But the repeatable feature of a propositional act is the propositional use of a sentence. The propositional use is a relation, and it is properties and relations which are repeatable in the strictest sense—as opposed to individuals which can be *copied.* However, it is misleading to assign truth and falsity directly to propositional uses, and it is misleading to attribute an eternal status to propositional uses just because their availability outruns their exemplifications. These are misleading because they distort the ontological status of propositional uses. Propositional uses as possibilities are rooted in the real world. They are possibilities for exemplification by real individuals. Its propositional use accounts for the truth of a sentence exemplifying it rather than being true in advance of its exemplification. The "content" of a propositional act is the proposition as exemplified. The sentence isn't true or false, and the sentence isn't the content of the act. But the sentence used in a certain manner is true or false, and as used it exemplifies, as one term of the relation, a propositional use.

It is easy to confuse what we can say now about the past and future with what was or will be true then. It is true now that there were nine planets before there was anyone around to say so. But it is misleading to claim that it was true then that there were nine planets. It is true *now* that there were nine planets *then.* The only sense in which a proposition could be true, though unexpressed, in the past is this: if someone had been there and had used a sentence in the appropriate manner, his sentence as so used would have been true. This much is true now. Just as we can say now that if some people in the past had been contestants in a race, then the contestant who crossed the finish line first without cheating or being otherwise disqualified would have been the winner. In fact, the contestant in a race who crosses the finish line first without cheating or being otherwise disqualified is *always* the winner. But so far no one has hypostatized eternal winners alongside the eternal propositions.

PART II BASIC AND COMBINATORIAL LOGIC

7: RICHARD J. ORGASS

Extended Basic Logic and Ordinal Numbers

Frederic Fitch has expressed informally some severe reservations about the theory of transfinite ordinals. In particular, he has expressed concern about its nonconstructive character. The purpose of this essay is to convince him that a substantial fragment of the theory of ordinal numbers can nonetheless prove acceptable to his point of view. This is done by showing that two different representations of the constructible ordinals are available in his system **K**′ of extended basic logic. For one of these representations, it is shown that transfinite induction to constructible ordinals is a derived rule of **K**′.

1. Introduction

1.1. It is assumed that the reader is familiar with the definition of the system **K** of basic logic [5, 8, 11, 12] and the system **K**′ of extended basic logic as described by Fitch. The material in section 2 uses a definition of **K** and **K**′ that is similar to the definitions given by Hermes [15].

1.2. The details of the proofs of some of the theorems in section 3 are greatly simplified if the theory of natural numbers of the system **R** of combinatory logic [19, 21] is available in **K**′. The relation of identity of the system **R** is defined in section 2, and it is shown that this relation is completely represented in **K**′. This means that the theory of natural numbers of the system **R** is available in **K**′.

1.3. The concept of a **U**-expression representing an ordinal is introduced in section 3. It is shown that each Church-Kleene ordinal [3] is represented by a **U**-expression that completely represents the set of **U**-expressions that represent ordinals.

1.4. Schütte [22] introduces ordinals by defining another order relation on the natural numbers. It is easy to establish connections between constructive ordinals and the natural numbers in Schütte's ordering. In section 4 it is shown that Schütte's order relation is completely represented in $\mathbf{K}'$, and that transfinite induction on these ordinals is a derived rule of $\mathbf{K}'$. Many of the proof-theoretic results in Schütte's book, *Proof Theory* [22], are established by transfinite induction on these ordinals. This result and Fitch's results concerning the representation of calculi [5, 6, 7, 11] in **K** and $\mathbf{K}'$ lead to the conclusion that $\mathbf{K}'$ provides a consistent meta language for Schütt's proof theory.

1.5. The notational conventions used in this essay follow the *Dictionary of Symbols of Mathematical Logic* [4].

2. *Natural Numbers*

2.1. The system **R** of combinatory logic was defined in [19, 21]. The relation of identity of the system **R** may be thought of as a relation which relates two **R**-formulas (wffs of the system **R**), just in the case that they are names for the same object. In 3.2 of [21], this relation is defined by giving an inductive definition of a set Δ of identities; definition 4.3, below, is equivalent. This section contains a summary of the proof of the statement that this relation is completely represented in $\mathbf{K}'$. A detailed proof is given in [20].

2.2. The symbol '$\doteqdot$' is used as a metalinguistic name for a relation among members of the class **U** of wffs of **K** and $\mathbf{K}'$. This relation is the same as the relation of identity of the system **R**. In order to avoid confusion, the relaton named by '$\doteqdot$' will be called *equality*.

2.3. *Definition.* Let 'K' serve as an abbreviation for '$\lambda_2 xy(x)$' and let 'S' serve as an abbreviation for '$\lambda_3 xyz(xz(yz))$'. The *relation $\doteqdot$ of equality* among **U**-expressions is defined inductively as follows: (1) 'a' $\doteqdot$ 'a'; (2) 'Kab $\doteqdot$ 'a'; (3) 'Sabc' $\doteqdot$ '$ac(bc)$'; (4) if 'b' is the result of substituting 'c' for one or more occurrences of 'd' in 'a', and if 'c $\doteqdot$ 'd', then 'a' $\doteqdot$ 'b'; (5) the only wffs that are equal to those which are equal by virtue of (1) to (4).

2.4. In order to show that there is a wff 'q' which represents $\doteqdot$ in **K**, it is useful to show that two relations among wffs are completely represented in **K**.

2.5. *Lemma.* There is a wff 'I' such that 'Iab' is in **K** (is in $\mathbf{K}'$) if 'a' occurs in 'b' and '$\sim$(Iab)' is in **K** (is in $\mathbf{K}'$) if 'a' does not occur in 'b'.

Proof. Let 'I' serve as an abbreviation for a wff such that $\mathrm{I}ab \leftrightarrow$ $[[a = b] \vee (\exists xy)\ [[b = xy]\ \&\ [\mathrm{I}ax \vee \mathrm{I}ay]]]$.

2.6. *Lemma.* There is a wff 'T' such that '$Tabcd$' is in **K** (is in **K**′) if 'a' is the result of substituting 'b' for an occurrence of 'c' in 'd' and otherwise '~($Tabcd$)' is in **K** (is in **K**′).
Proof. Let 'T' serve as an abbreviation for a wff such that $Tabcd \leftrightarrow [[[b = c] \vee \sim(Icd] \;\&\; [a = d]] \vee [[c = d] \;\&\; [a = b]] \vee (\exists xy)[[d = xy] \;\&\; (\exists a_1 a_2)\,[[a = a_1 a_2] \;\&\; Ta_1 bcx \;\&\; Ta_2 bcy]]]$.
2.7. *Theorem.* There is a wff 'q' that represents in **K** the relation ≑ of equality among wffs.
Proof. Let 'q' serve as an abbreviation for a wff such that: $qab \leftrightarrow [[a = b] \vee (\exists c)\,[a = Kbc] \vee (\exists cdx)\,[[a = Scdx \;\&\; [b = cx(dx)]] \vee (\exists cd)\,[Tbcda \;\&\; qcd]]$.
2.8. *Theorem.* There is a wff 'Q' that completely represents in **K**′ the relation ≑ of equality among wffs.
Proof. By [7] there is a wff 'k' that completely represents in **K**′ the class **K** of wffs. Let 'Q' serve as an abbreviation for a wff such that $Qab \leftrightarrow k(qab)$. Hereafter, let '≑' serve as an abbreviation for 'Q'.
2.9. Since the relation of equality of the system **R** is completely represented in **K**′, the theory of this relation in the system **R** is available in **K**′. In particular, the theorems concerning the representation of natural numbers and functions of natural numbers in the system **R** are available in **K**′. This will greatly simplify the discussion in sections 4 and 5 below. The remainder of this section contains a summary of some of these properties. Proofs of these statements which apply in **K**′ for the system **R** are given in [19, 21].
2.10. Define a countable set of relations of identity among **U**-expressions as $a \doteqdot_1 b \leftrightarrow (x)\,[ax \doteqdot bx]$, and $a =_2 b \leftrightarrow (x,y)\,[axy \doteqdot bxy]$, and so forth. Obviously, each of these relations is completely represented in **K**′. A **U**-expression 'a' is said to represent the natural number n if

$$axy \doteqdot \underbrace{x(x(x \ldots (x}_{n \text{ times}}y) \ldots))$$

is in **K**′. Each natural number is represented by an **R**-numeral; let '0', '1', '2', and so on serve as abbreviations for these **U**-expressions. There is a **U**-expression 'N' which completely represents the set of **U**-expressions which represent natural numbers. A **U**-expression 'f' is said to represent an n-ary function f in K′ iff

$fa_1 \ldots a_n =_2 a$ is in $\mathbf{K}'$, iff $f(x_1, \ldots, x_n) = x$ where 'a_1', 'a_2', . . ., 'a_n' and 'a' respectively represent $x_1, x_2, \ldots, x_n$ and x. Each partial recursive function is represented by a **U**-expression. A **U**-expression 'a' is said to *completely represent* a relation A among natural numbers iff 'axy' is in $\mathbf{K}'$, if A(x,y) and '$\sim(axy)$' is in $\mathbf{K}'$ if not A(x,y) where 'x' and 'y' represent x and y respectively. Each partial recursive relation among natural numbers has a complete representation in $\mathbf{K}'$.

2.11. The abstraction rules of **R** are available in $\mathbf{K}'$. For all $n \geqslant 1$, there is an effective procedure to define '$\Lambda_n x_1 \ldots x_n$ $(---x_1, \ldots, x_n ---)$', so that $\Lambda_n x_1 \ldots x_n(---x_1, \ldots, x_n ---)$ $a_1 \ldots a_n \doteq (---a_1, \ldots, a_n ---)$ is in $\mathbf{K}'$, and the rule of equality elimination applies to **U**-expressions and subexpressions of the form '$\doteq$ab'. This means that the obvious analog of Church's λ-K-calculus in [3] is available in $\mathbf{K}'$.

3. Church-Kleene Ordinals in K'

3.1. The ordinals in the *first number class* are the finite ordinals: **0**, **1**, **2**, **3**, . . . The limit of this sequence of ordinals, called ω, is the first ordinal in the *second number class.* An ordinal which is the limit of a sequence of ordinals is said to be a *limit* ordinal; ω is the first limit ordinal. If we are speaking in terms of von Neumann ordinals, ω is the set of all finite ordinals. Fitch has shown that each natural number has a representation in $\mathbf{K}'$. Further, there is a **U**-expression in 'N' that completely represents the set of **U**-numerals. Consequently, $\mathbf{K}'$ contains the ordinal ω in a way which will be made precise below.

3.2. Church and Kleene [1, 3] use the term formally definable ordinal to refer to ordinals which are represented by λ-expressions. λ-expressions represent ordinals in very much the same way that Church's λ-numerals represent natural numbers. The λ-expressions which represent limit ordinals are defined in terms of the order type of the sequence and the function used to define the sequence. They are able to provide λ-expressions which represent the ordinals in the second number class. Further, they provide a constructive proof that their set of ordinals is simply ordered. It is possibly closely to parallel their proofs in $\mathbf{K}'$.

3.3. Church and Kleene use the following abbreviations to define their ordinals: '**0**' for '$\Lambda a(a1)$', 'S_o' for '$\Lambda_2 ab(b2a)$', and 'L' for '$\Lambda_3 abc(c3ab)$'. '**1**', '**2**', '**3**', . . . , respectively, are abbreviations for '$S_o 0$', '$S_o 1$', '$S_o 2$', . . . When necessary to avoid confusion, the subscript 'o' is used to refer to ordinals and functions of ordinals. Here is how these **U**-expressions represent finite ordinals: **0**$b \doteqdot b$1, **1**$b \doteqdot b$2**0**, **2**$b \doteqdot b$21, and **3**$b \doteqdot b$22. Observe that the representing formula contains a coding of just how the ordinal was constructed using the successor function. The following definition of a **U**-expression representing an ordinal is a modification of the Church-Kleene definition.

3.4. *Definition.* A **U**-expression is said to *represent an ordinal* just in the case that it can be shown to represent an ordinal by virtue of one of the following rules: (1) if 'a' represents the ordinal α and $a =_I b$, then 'b' also represents α; (2) '**0**' represents the ordinal zero; (3) if 'a' represents the ordinal α, then '$S_o a$' represents the successor of α; (4) if α is the limit of an increasing sequence of ordinals, $\alpha_0, \alpha_1, \alpha_2, \ldots$ of order type ω and if 'r' is a **U**-expression such that the **U**-expressions 'r**0**', 'r**1**', 'r**2**', . . . represent the ordinals $\alpha_0, \alpha_1, \alpha_2, \ldots$, respectively, then 'L**0**r' represents α. An ordinal of the first or second number class is said to be **K**$'$ *definable* if there is a **U**-expression that represents the ordinal. If a **U**-expression represents an ordinal, it is said to be a **U**-ordinal.

3.5. The least ordinal ω of the second number class is the limit of the sequence **0**, **1**, **2**, . . . By definition 3.4 (4). 'L**0**I' represents the ordinal ω. Let 'ω' serve as an abbreviation for 'L**0**I'. The following illustrates how 'ω' represents the set of finite ordinals:

$$
\begin{aligned}
\omega\mathbf{0} &\doteqdot \mathrm{L}\mathbf{0}\mathrm{I}\mathbf{0} \\
&\doteqdot 3\mathbf{1}\mathbf{0}\mathrm{I} \\
\omega\mathbf{1} &\doteqdot \mathrm{L}\mathbf{0}\mathrm{I}\mathbf{1} \\
&\doteqdot 3200\mathrm{I} \\
\omega\mathbf{2} &\doteqdot \mathrm{L}\mathbf{0}\mathrm{I}\mathbf{2} \\
&\doteqdot 3210\mathrm{I}.
\end{aligned}
$$

Notice that the concept of a **U**-expression representing an ordinal differs from the concept of a **U**-expression representing a set of **U**-expressions. Also, observe that 'ω' is defined in such a way that the **U**-expression 'ωa', where 'a' represents a finite ordinal α, contains a description of how α was built up from **0** using the successor function and the identity function and of how ω is the limit of a sequence of ordinals, namely, **0**, **1**, **2**, **3**, . . .

3.6. *Definition* (Church-Kleene [3]). A sequence of ordinals of order type ω is **K**′ *defined as a function of ordinals by* ‘*r*’ if, for every **U**-ordinal ‘*a*’ which represents a finite ordinal α, ‘*ra*’ represents the $(1 + \alpha)$th ordinal of the sequence.

3.7. If the definition of normal form for the λ-calculus is modified in the obvious way to deal with the abstraction of **R** that is available in **K**′, the Church-Kleene constructive proofs of the following theorems apply to **U**-ordinals.

3.8. *Theorem* (Church-Kleene [3]). Every **U**-ordinal has a normal form.

3.9. *Theorem* (Church-Kleene [3]). If a **U**-ordinal ‘*a*’ represents an ordinal α, then ‘*a*’ cannot represent an ordinal distinct from α.

3.10. In order to consider the question of which ordinals are represented by **U**-ordinals, it is necessary to define the concept of a **U**-expression representing a function of ordinals. The following is a minor modification of the Church-Kleene definition.

3.11. *Definition.* A function is said to be a *function in the first and second number class* if the domain and range of the function are ordinals in the first and second number class. A **U**-expression ‘f’ is said to *represent the n-ary function f* in the first and second number class if ‘$\mathrm{f}a_1 \ldots a_n$’ is a **U**-ordinal and if ‘$\mathrm{f}a_1 \ldots a_n =_1 a$’ is in **K**′ just in the case that $f(\alpha_1, \ldots, \alpha_n) = \alpha$ where ‘a_1’, . . . , (a_n’ and ‘*a*’, respectively, represent the ordinals $\alpha_1, \ldots, \alpha_n$ and α.

3.12. The next two paragraphs contain an outline of the essential details of the proof of the statement that all ordinals which are definable as limits of sequences of order type ω are represented by **U**-ordinals. In particular, the functions (ordinal) addition, multiplication, and exponentiation are represented in **K**′. In this discussion, it will be claimed that there are **U**-expressions with certain properties. The existence of these **U**-expressions is a direct consequence of:

3.13. *Theorem* (Church-Kleene [3]). If ‘*a*’, ‘*b*’, and ‘*c*’ are **U**-expressions built up out of ‘S’ and ‘K’, it is possible to find eight **U**-expressions ‘f_{ijk}’ (where the subscripts *i, j, k* take the values *1* and *2*) with the following properties: $\mathrm{f}_{ijk}\mathbf{0} =_n a$, $\mathrm{f}_{2jk}\mathbf{0} =_n a\,\mathrm{f}_{1jk}$, $\mathrm{f}_{i1k}(\mathrm{S}_0 a) =_n ba$, $\mathrm{f}_{i2k}(\mathrm{S}_0 a) =_n c\,\mathrm{f}_{i2k}\,a$, $\mathrm{f}_{ij1}(\mathrm{L}ar) =_n car$, and $\mathrm{f}_{ij2}(\mathrm{L}ar) =_n c\,\mathrm{f}_{ij2}\,ar$. The subscript *n* in the sign ‘$=_n$’ depends on the **U**-expressions *a, b,* and *c* only.

Proof. The proof of this theorem is simply a modification of the proof of theorem 3 of [3]. The proof is related to the Church-

Kleene proof in the same way that the proof of theorem 7.6 of Chapter 2 of [19] is related to Church's proof that primitive recursive functions are λ-definable [21].

3.14. There is a **U**-expression 'f' such that: f**0**$b =_I$ I$b =_I b$, f($S_0 a$)$b =_I S_0$ (fab), and f(Lar)$b =_I$ La(Λm(f(rm)b)). It will now be shown that if 'a' and 'b', respectively, represent ordinals α and β, then 'fab' represents the sum of α and β. If α is not a limit ordinal, then the first two properties of 'f' can be used to show:

$$\text{f}ab =_I \underbrace{S_0\,(S_0 \ldots (S_0}_{\gamma \text{ times}}\,(\text{f}db) \ldots),$$

where 'd' represents the largest limit ordinal δ less than α and $\alpha = \delta + \gamma$. At this point, the third property of 'f' is applicable. This is precisely what is required. Now suppose α is a limit ordinal. As an example, consider ω, the limit of the sequence **0**, **1**, **2**, **3**, . . . In this case, 'a' is 'L**0**I', so the third property of 'f' applies: f(L**0**I)$b =_I$ L**0**(Λm(fIm))b. That is, by definition 3.4 (4), the result is the limit of the sequence β, β + **1**, β + **2**, . . . A similar argument applies if 'a' represents some other limit ordinal. Therefore 'f' represents ordinal addition. Let '[$b +_0 a$]' serve as an abbreviation for 'fab'.

3.15. By a similar modification of the primitive recursive definitions, it is possible to exhibit **U**-expressions which represent multiplication, exponentiation, and the constant function. Let '$\times_0$', 'exp_0', and 'K_0' serve as abbreviations for these **U**-expressions. Further, let 'b^a' serve as an abbreviation for '$exp_0\ ba$'. The statements in 3.14 and 3.15 are the essential details of the proof of:

3.16. *Theorem.* Each of the finite ordinals and each ordinal definable by addition, multiplication, exponentiation, and as the limit of a sequence of order type ω has a representation in **K**$'$.

Proof. If an ordinal δ is the sum, product, or exponentiation of two ordinals α and β, then by theorem 3.13 and 3.14 and 3.15 there is a **U**-ordinal that represents δ. Similarly, if δ is the limit of the sequence $f(\alpha,\mathbf{0}), f(\alpha,\mathbf{1}), f(\alpha,\mathbf{2})$, . . . then, by definition 3.4(4), 'Lfa' represents δ where 'f' represents f and 'a' represents α.

3.17. By theorem 3.16, there are **U**-expressions which represent the epsilon numbers. There is another more direct way of showing that these ordinals are represented by **U**-ordinals. The epsilon numbers are the fixed points of the continuous, monotonically increasing function f defined by the identity $f(\alpha) = \omega^\alpha$. It will now be shown that there are **U**-ordinals which represent the fixed points of f and that these **U**-expressions also represent the limits of the appropriate sequences. A number of preliminary results, which are modifications of results of Church and Kleene [3] are required. The proofs closely follow the proofs in [3], so they are omitted.

3.18. *Lemma.* There is a **U**-expression 'p' such that $\mathrm{P}\mathbf{0} =_1 \mathbf{0}$, $\mathrm{P}(\mathrm{S_o}a) =_1 a$, and $\mathrm{P}(\mathrm{L}ar) =_1 \mathrm{L}ar$. 'P' represents the predecessor function on ordinals. (The predecessor of **0** is **0**, the predecessor of a nonlimit ordinal is the ordinal that precedes it, the predecessor of a limit ordinal is the limit ordinal.)

3.19. *Lemma.* There is a **U**-expression 'f' such that $\mathrm{f}\mathbf{0} =_1 \mathbf{0}$, $\mathrm{f}(\mathrm{S_o}a) =_1 a$ and $(\mathrm{fL}ar) =_1 \mathrm{K_o}\mathbf{1}(\mathrm{L}ar) =_1 1$. 'f' represents a function from ordinals to ordinals defined as follows:

$$f(\alpha) = \begin{cases} \mathbf{0} \text{ if } \alpha \text{ is a finite ordinal;} \\ \mathbf{1} \text{ if } \alpha \text{ is an infinite ordinal.} \end{cases}$$

3.20 *Lemma.* There is a **U**-expression 't' such that $\mathrm{t}\mathbf{0} =_1 \mathbf{0}$, $\mathrm{t}(\mathrm{S_o}a) =_1 \mathrm{K_o}\mathbf{1}a =_1 \mathbf{1}$ and $\mathrm{t}(\mathrm{L}ar) =_1 \mathrm{K_o}\mathbf{2}a =_1 \mathbf{2}$. 't' represents the function t:

$$t(\alpha) = \begin{cases} \mathbf{0} \text{ if } \alpha \text{ is the ordinal } \mathbf{0}; \\ \mathbf{1} \text{ if } \alpha \text{ is a successor ordinal;} \\ \mathbf{2} \text{ if } \alpha \text{ is a limit ordinal.} \end{cases}$$

3.21. *Lemma.* The **U**-expression '$\Lambda x(\mathrm{P}(x\mathrm{S_o}\mathbf{0}))$' represents (using the appropriate definition) a function from natural numbers to finite ordinals such that its value for argument n is the nth finite ordinal. There is a **U**-expression 'T' such that $\mathrm{T}\mathbf{0} =_2 1$, $\mathrm{T}(\mathrm{S_o}a =_2 +1(\mathrm{T}a)$, and $\mathrm{T}(\mathrm{L}ar) =_2 \mathrm{K}1(\mathrm{L}ar)$. 'T' represents a function T from ordinals to natural numbers with the following properties. If α is a finite ordinal, then $T(\alpha)$ is the αth natural number. If α is an infinite ordinal, then $T(\alpha)$ is the integer n such that if β is the largest ordinal less than or equal to α, then $\alpha = S_o{}^n(\beta)$.

3.22. *Theorem.* There is a **U**-expression 'e' such that if a is a **U**-ordinal which represents α, then '$\mathrm{e}a$' is a **U**-ordinal which represents the ordinal which is the $(\alpha + \mathbf{1})$st fixed point of the function f defined by the identity $f(\alpha) = \omega^\alpha$. Furthermore, '$\mathrm{e}\mathbf{0}$' represents the limit of the sequence

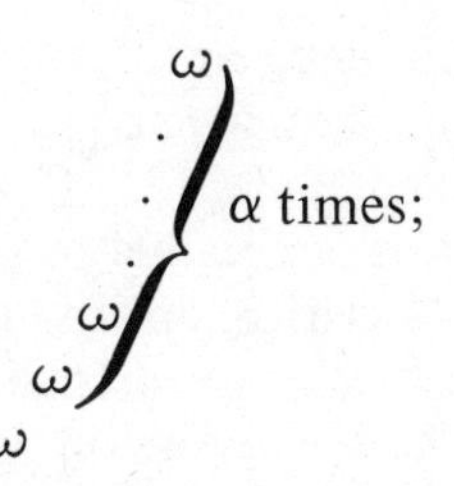

'e1' represents the limit of the sequence

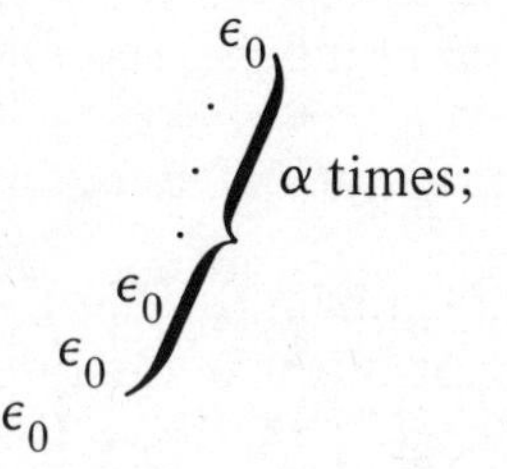

and, in general, '$e(S_o a)$', where 'a' represents the ordinal α, is the limit of

$$\epsilon_\alpha^{\epsilon_\alpha^{\epsilon_\alpha^{\cdot^{\cdot^{\cdot}}}}}$$

Proof. It is straightforward to modify the proof given by Church and Kleene [3] to show that there is a **U**-expression 'e' with the following properties: $e\mathbf{0} =_1 \mathrm{L}\mathbf{0}\Lambda x(\mathrm{T}x(exp_o\ \omega)\mathbf{0})$, $e(S_o a) =_1 \mathrm{L}\mathbf{0}\Lambda x(\mathrm{T}x(exp_o\ \omega)(S_o(ea)))$, and $e(\mathrm{L}ar) =_1 \mathrm{L}a\Lambda x(e(rx))$. By definition 6.2(4) and Lemma 6.19, '$e\mathbf{0}$' represents the limit of the sequence $\omega, \omega^\omega, \omega^{\omega^\omega}, \ldots$, i.e., ϵ_0. If 'a' represents a successor ordinal, then 'ea' represents the limit of

$$\cdot^{\cdot^{\cdot}}$$

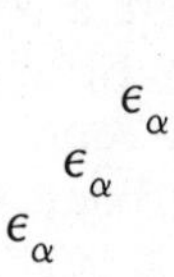

where 'a' represents the successor of α for the same reason. Finally, if 'a' represents the limit of the sequence $r(\beta,0), r(\beta,1)\ r(\beta,2)$, . . . , then '$ea$' represents the limit of the sequence $\epsilon_{r(\beta,0)}, \epsilon_{r(\beta,1)}, \epsilon_{r(\beta,2)}, \ldots$

3.23. In a completely analogous way, it is possible to show that there are **U**-ordinals which represent the critical epsilon numbers. In particular, in the definition of 'e' in the proof of theorem 3.22, replace occurrences of '$exp_0\ \omega$' with the **U**-expression which represents the function f defined by the identity $f(\alpha) = \epsilon_\alpha$. By theorem 3.22, there is such a **U**-expression.

3.24. *Theorem.* Each critical epsilon number is represented by a **U**-ordinal.

3.25. *Theorem.* There is a **U**-expression '*Ord*' that completely represents the set of **U**-ordinals.

Proof. Definition 3.4's (1), (2), and (3) obviously define a recursive set of **U**-expressions. This is the set of those **U**-expressions which represent finite ordinals, and by [7] there is a **U**-expression 'O' that completely represents this set. Now, use the function T of lemma 6.19 to express the order condition of 3.4 (4) in $\mathbf{K}'$, $\mathfrak{L}r \leftrightarrow (x)\ [\mathrm{O}x \supset \mathrm{T}\ (rx) < \mathrm{T}(r(\mathrm{S}_0 x))]$, where '<' completely represents the relation 'less than' among natural numbers. By the closure rules for excluded middle, '$\mathfrak{L}$' is definite. In fact, $\mathfrak{L}$ is primitive recursive. Use Fitch's procedure for defining self-referential relations [9] to define a **U**-expression '*Ord*' with the following property: $Ord\ a \leftrightarrow a = \mathbf{0} \lor (\exists b)\ [Ord\ b\ \&\ a =_1 b] \lor (\exists b)\ [Ord\ b\ \&\ a =_1 \mathrm{S}_0 b] \lor (\exists r)\ [\mathfrak{L}r\ \&\ a =_1 \mathrm{L}\mathbf{0}r]$. Note that each of the existentially quantified subexpressions are definite and that the remaining clause uses the identity of $\mathbf{K}'$. Consequently, by induction on finite ordinals it can be shown that for all **U**-expressions 'a', either '$Ord\ a$' or '$\sim(Ord\ a)$' (but not both) is in $\mathbf{K}'$. Therefore, '*Ord*' completely represents the set of **U**-ordinals.

3.26. It has been shown that all ordinals which are definable by means of the successor function, addition, multiplication, and exponention as well as ordinals which are limits of sequences of order type ω are represented by **U**-ordinals. Further, the set of **U**-ordinals is completely represented in $\mathbf{K}'$. It is possible to continue this development and show that the binary relation 'less than' which well-orders the **U**-ordinals is completely represented in $\mathbf{K}'$. A proof of this result is given for Schütte's ordinals in section 4.

4. Schütte Ordinals in K′

4.1. Schütte [22] develops a theory of constructible ordinals by defining a decidable order relation on the natural numbers. This well-ordering has the property that prime numbers and prime numbers with prime exponents correspond to limit ordinals. For example, 3 corresponds to ω, 3^3 corresponds to ω^ω, and 5 corresponds to ϵ_0. Schütte's ordinals as well as the set of Schütte ordinals are clearly represented by **U**-expressions. In this section, it is shown that Schütte's ordinals are well ordered in **K**′ and that transfinite induction on these ordinals is a derived rule of **K**′.

4.2. As was shown in section 2, the arithmetic of natural numbers developed in the system R is available in **K**′, and the results of [19] will be used when required. 'p_0', 'p_1', 'p_2', . . . are used as names for the prime numbers $2, 3, 5, \ldots$

4.3. After some technical preliminaries, Schütte's definition of two binary relations under ($\subset$) and before ($\prec$), on natural numbers is stated in 4.7. Paragraph 4.11 contains a proof that these relations are completely represented in **K**. After this, it is shown that $\prec$ well-orders the natural numbers in **K**′. Then in 4.29 it is shown that transfinite induction for the relation $\prec$ is a derived rule of **K**′.

4.4. If i is a natural number, then i_j is the exponent of the jth prime in the (unique) prime decomposition of i, e.g., $i = p_0{}^{i_0} \cdot p_1{}^{i_1} \cdot \ldots \cdot p_n{}^{i_n}$. It is well known that there exists a k such that for all $j > k$, $i_j = 0$. Continuing this definition, i_{jk} is the exponent of the kth prime in the prime decomposition of i_j, e.g. $i_j = p_0{}^{i_{j0}} \cdot p_1{}^{i_{j1}} \cdot \ldots \cdot p_m{}^{i_{jm}}$. This convention is extended to any finite number of subscripts. For all natural numbers $i, j, k, n,$ and so on, $i > i_j > i_{jk} > i_{jkn} > \ldots \geqslant 0$. Numbers such as i_j, i_{jk}, and i_{jkn} are called *unary, binary,* and *ternary indices* of i, respectively. A similar statement applies for more than three subscripts.

4.5. *Lemma.* There is a **U**-expression 'ℓ' which represents the function ℓ whose value for argument i is the integer k such that $i_j = 0$ for all $j > k$. There is a **U**-expression '$G\ell$' which represents the binary function Gℓ defined by the identity $G\ell(i,j) = i_j$, provided $j \leqslant \ell(i)$. Further, for all $n \geqslant 1$, there is a function which maps an interger onto its n-ary indices; these functions are represented by **U**-expressions.

Proof. Gödel [13] has given primitive recursive definitions of ℓ

and $G\ell$. By 2.10, there are **U**-expressions which represent these functions. Let the function ι be defined as follows:

$$\iota(i,j) = \begin{cases} G\ell(i,j), \text{ if } j \leqslant \ell(i); \\ 0, \text{ if } j > \ell(i). \end{cases}$$

Since ℓ and $G\ell$ are primitive recursive functions, ι is a primitive recursive function [14]. Therefore, there is a **U**-expression which represents ι. For binary subscripts, $\iota_2(i,j,k) = \iota(\iota(i,j),k)$ is the appropriate function. A similar argument applies for n-ary indices. Hereafter, straightforward proofs of this kind will be omitted.

4.6. *Definition.* Stated informally, the *rank* of a natural number i is the largest n such that there is a nonzero n-ary index of i. The *rank* $\rho(i)$ of a natural number is defined recursively as follows:

$$\rho(i) = \begin{cases} 0 & \text{if } i = 0; \\ \left[\max\limits_{0 \leqslant j \leqslant \ell(i)} \rho(i_j)\right] + 1, & \text{if } i \neq 0 \end{cases}$$

Since $\rho(i) > \rho(i_j)$, this is a recursive definition of ρ and, therefore, there is a **U**-expression 'ρ' which represents ρ. For example, only 0 has rank 0, and 1 is the only natural number with rank 1; only primes and pairwise products of primes have rank 2.

4.7. *Definition* (Schütte [22]). The binary relations under ($\subset$) and before ($\prec$) are defined by simultaneous induction as follows: (1) if $i \neq 0$, then $0 \subset i$ and $0 \prec i$; (2) if $i \neq 0$ and $j \neq 0$, then $i \subset j$ if there exists a k such that $i_k \prec j_k$, and for all $n > k$, $i_n = j_n$; (3) if $i \neq 0$ and $j \neq 0$, then $i \prec j$ if for all $k < \ell(i)$, $i_k \prec j$; and (4) if $i \neq 0$ and $j \neq 0$, then $i \prec j$ if $j \subset i$ and if for at least one k, $i \prec j_k$.

4.8. Some illustrations of the properties of $\prec$ and $\subset$ are useful at this point. By 4.7(1), $0 \prec i$ and $0 \subset i$ if $i \neq 0$. By 4.7(2), $1 \subset i$ for all $i > 1$. By 4.7(3), $1 \prec i$ for all $i > 1$. It can be shown that the order relation $\prec$ is such that the successor function σ is defined by the identity $\sigma(i) = 2^i$. The initial segment of this ordering is $0, 1, 2, 2^2, 2^{(2^2)}, \ldots$ (Hereafter, association right is used for exponents, so that 2^{2^2} is an abbreviation for $2^{(2^2)}$.) 4.7(2) and (3) can be used to show $0 < 1 < 2 < 2^2 < 2^{2^2} < \ldots$ Furthermore, for any n,

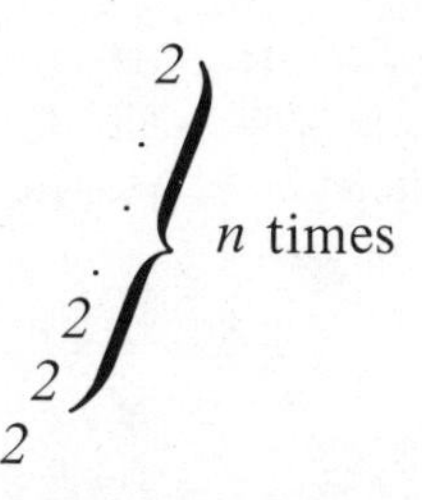

(*)

is before *3* by 4.7(2) and (3). By 4.7(2) and (4), $3 \prec 2^m$ where m is a prime. By a similar argument, $3 \prec 2^m$ if m is not of the form (*). Consequently, $0, 1, 2, 2^2, 2^{2^2}, \ldots$ are the finite ordinals, and *3* is ω.

4.9. The successor of *3* is 2^3, and the ordering continues $3, 2^3, 2^{2^3}, \ldots$ Using definition 4.7 it can be shown that only natural numbers of the form

$$2^{2^{2^{\cdot^{\cdot^{\cdot^{2^{3}}}}}}} \qquad (**)$$

are after *3* and before 3^2. Thus, this sequence corresponds to ω, $\omega + 1, \omega + 2, \ldots$, and 3^2 corresponds to $\omega + \omega$ or $\omega \times 2$. Continuing with the successor function, we get natural numbers of the form

Each of these numbers has the property that it is after 3^2 and before 3^{2^2}. In addition, there are no other integers after 3^2 and before 3^{2^2}. Therefore, 3^{2^2} corresponds to $\omega \times 3$. It is easy to see

that $3, 3^2, 3^{2^2}, 3^{2^{2^2}}, \ldots$ corresponds to the ordinals ω, $\omega \times 2$ $\omega \times 3$, $\omega \times 4, \ldots$, respectively. Continuing with a similar argument, it can be shown that 3^3 corresponds to ω^ω. Further, 5 is after

$$3^{3^{3^{\cdot^{\cdot^{\cdot}}}}},$$

so 5 corresponds to ϵ_0.

4.10. The relation $\prec$ has the property that $2^i \cdot 5$ where $i \prec 2^i \cdot 5$ is the ith epsilon number. $3 \cdot 5$ corresponds to χ_0, the first critical epsilon number, and $2^i \cdot 3 \cdot 5$, where $i \prec 2^i \cdot 3 \cdot 5$ corresponds to χ_i, the ith critical epsilon number.

4.11. *Theorem.* There are **U**-expressions '$\subset$' and '$\prec$' which completely represent the relations $\subset$ and $\prec$ in **K**$'$.

Proof. Using complete induction on the sum of the ranks of i and j, Schütte [22, theorem 11.1], proves that $\prec$ and $\subset$ are effectively decidable. Therefore, there is a (total) recursive function $f_\subset$ such that

$$f_\subset(i,j) = \begin{cases} 0, \text{ if } i \prec j; \\ 1 \text{ otherwise.} \end{cases}$$

By 2.10, there is a **U**-expression '$\mathrm{f}_\subset$' which represents $f_\subset$. Let '$\subset$' serve as an abbreviation for '$\lambda_2 xy[\mathrm{f}_\subset iy =_2 0]$'. Then '$\subset$' completely represents the relation 'under'. A similar argument applies for '$\prec$'.

4.12. In 4.13 to 4.28 it is shown that induction on the natural numbers ordered by the relations $\prec$ and $\subset$ is available in **K**$'$. Induction for the relation $\subset$ is established in the metalanguage only because it is needed to establish induction for $\prec$ in **K**$'$. These proofs are completed by induction on natural numbers (in their usual order.)

4.13. *Lemma.* For all natural numbers i, '$\sim[\mathrm{i} \subset \mathrm{i}]$' and '$\sim[\mathrm{i} \prec \mathrm{i}]$' are in **K**$'$ (antireflexive).

Proof. The proof is by induction on the rank of i. If $\rho(i) = 0$, then, by definition 4.6, $i = 0$. By definition 4.7, it is not the case that $i \subset i$ or $i \prec i$. Therefore, by theorem 4.11, '~[i ⊂ i]' and '~[i ≺ i]' are in **K**′.

The inductive step is to set $\rho(i) = n > 0$. Thus, $i \neq 0$. The inductive hypothesis is that for all j such that $\rho(j) < n$, '~[j ⊂ j]' and '~[j ≺ j]' are in **K**′. Only 4.7(2) can be used to show that $i \subset i$. Therefore, there must be a k such that '~[$i_k \prec i_k$]' is in **K**′, and $\rho(i_k) < \rho(i)$. This contradicts the inductive hypothesis so, by theorem 4.11, '~[i ⊂ i]' is in **K**′. Only 4.7(3) and (4) can be used to show that $i \prec i$. Both of these clauses require that 'i ⊂ i' is in **K**′. This has already been shown to be false. Therefore, by theorem 4.11, '~[i ≺ i]' is in **K**′. This argument can be carried out entirely inside **K**′.

4.14. *Lemma.* If '~[i $=_2$ j]' is in **K**′, then either 'i ⊂ j' or 'j ⊂ i' and either 'i ≺ j' or 'j ≺ i' are in **K**′ (totality).

Proof. The proof is by induction on the sum of the ranks of i and j. If $\rho(i) + \rho(j) = 0$, then $i = j$ so 'i $=_2$ j' is in **K**′. The inductive step is to let $\rho(i) + \rho(j) = n > 0$ and to assume the theorem for all i and j such that $\rho(i) + \rho(j) < n$. If '~[i $=_2$ j]' and 'i $=_2$ 0' are in **K**′, then by definition 4.7 and theorem 4.11, 'i ⊂ j' and 'i ≺ j' are in **K**′. A similar argument applies if '~[i $=_2$ j]' and 'j $=_2$ 0' are in **K**′. The remaining case is '~[i = j]', '~[i = 0]', and '~[j = 0]' are in **K**′.

There are two subcases: (1) There is a k such that $i_k \neq j_k$ and $i_m = j_m$ for all $m > k$. $\rho(i_k) + \rho(j_k) < n$, so the lemma holds for i_k and j_k. Therefore, by definition 4.7 and theorem 4.11, 'i ⊂ j' or 'j ⊂ i' is in **K**′. (2) Without loss of generality, assume 'i ⊂ j' is in **K**′. Therefore, for all $k < \ell(i)$, '$i_k \prec$ j' is in **K**′. By definition 4.7, $i \prec j$; by theorem 4.11, 'i ≺ j' is in **K**′. In case 4.7(4) was used, there is a k such that not $i_k \prec j$. By the inductive hypothesis, the theorem holds for $j \prec i_k$. Therefore, by 4.7(4) $j \prec i$ and 'j ≺ i' is in **K**′ by theorem 4.11. This completes the proof.

4.15. *Lemma.* If 'i ⊂ j' and 'j ⊂ k' are in **K**′ then 'i ⊂ k' is in **K**′. If 'i ≺ j' and 'j ≺ k' are in **K**′, then 'i ≺ k' is in **K**′ (transitivity). (The proof which is a modification of Schütte's proof of theorem 11.4 [22] in the same way that the proof of lemma 4.14 is a modification of Schütte's proof of theorem 11.3, is omitted.)

4.16. *Definition* (Schütte [22], p. 112). A property A of natural numbers is said to be *≺-progressive* (*⊂-progressive*) if $A(i)$ for all

$i \prec j$ ($i \subset j$) implies $A(j)$. *Transfinite $\prec$-induction* (*$\subset$-induction*) asserts that every *$\prec$-progressive* (*$\subset$-progressive*) property holds for all natural numbers.

4.17. Here is an outline of the proof that transfinite $\prec$-induction is a derived rule of **K**$'$. Let I be a property of natural numbers which holds if $\prec$-induction holds in **K**$'$ to the natural number *i*. It will be shown that I is $\prec$-progressive. This result is used to show that transfinite $\prec$-induction is a derived rule of **K**$'$. Only induction on natural numbers (in their usual order) and the fact that $\prec$ and $\subset$ are completely represented in **K**$'$ are used.

4.18. *Convention.* In the metalanguage, an expression of the form $(\forall a \epsilon b)$ $[(\ldots a \ldots) \rightarrow (\text{- - -} a \text{ - - -})]$ will be used to express the statement "For all **U**-expressions '*a*' such that '*ba*' is in **K**$'$, if '$(\ldots a \ldots)$' is in **K**$'$ then '$(\text{- - -} a \text{ - - -})$' is in **K**$'$."

4.19. *Definition.* Let $P(a)$ be a metalinguistic abbreviation for $(\forall \mathrm{i} \epsilon \mathrm{N})$ $[(\forall \mathrm{j} \epsilon \mathrm{N})$ $[\mathrm{j} \prec \mathrm{i} \rightarrow a\mathrm{j}] \rightarrow a\mathrm{i}]$. That is, $P(a)$ asserts that '*a*' represents a property of natural numbers which is $\prec$-progressive in **K**$'$. Let '$\mathrm{I}'(a,\mathrm{i})$' serve as a metalinguistic abbreviation for: '*if* $P(a)$ *then* $(\forall \mathrm{j} \epsilon \mathrm{N})$ $[\mathrm{j} \prec \mathrm{i} \rightarrow a\mathrm{j}]$'. That is, $\mathrm{I}'(a,i)$ asserts that if '*a*' represents a progressive property of natural numbers in **K**$'$ then for all integers, *j*, before or equal to *i*, '*aj*' is in **K**$'$. Stated in another way, $\mathrm{I}'(a,i)$ asserts that transfinite $\prec$-induction holds in **K**$'$ for '*a*' up to '*i*'. Finally, let $\mathrm{I}(i)$ serve as a metalinguistic abbreviation for the statement: '$(\forall a \epsilon \mathbf{U})$ [*if* $P(a)$ *then* $(\forall \mathrm{j} \epsilon \mathrm{N})$ $[\mathrm{j} \prec \mathrm{i} \rightarrow a\mathrm{j}]$]'. That is, for the statement that for all **U**-expressions, '*a*', if '*a*' represents a progressive property of natural numbers in **K**$'$, then $\prec$-induction to *i* is a derived rule of **K**$'$.

4.20. *Lemma.* If $\prec$-induction to some integer *j* is a (derived) rule of **K**$'$, then it is a (derived) rule of **K**$'$ for any integer $i \prec j$. That is, if $\mathrm{I}(j)$ and $i \prec j$, then $\mathrm{I}(i)$. (This follows from the definition of I.)

4.21. *Lemma.* If $P(a)$, then '$a0$' is in **K**$'$.

Proof. Let $i = 0$ in the definition of P: $(\forall \mathrm{j} \epsilon \mathrm{N})$ $[\mathrm{j} \prec 0 \rightarrow a\mathrm{j}] \rightarrow a0$ There is no natural number *j* such that $j \prec 0$ by definition 4.7. Consequently, since $\prec$ is completely represented in **K**$'$, the antecedent is vacuously true and '$a0$' is in **K**$'$. (*Remark.* Lemmas 4.20 and 4.21 are also provable if they refer to a property, A, of natural numbers; Schütte provides the proof in [22], p. 104.)

4.22. *Lemma.* $\mathrm{I}(i)$ is $\prec$-progressive.

Proof. On the hypothesis that $\mathrm{I}(j)$ for all $j \prec i$, it will be shown that $\mathrm{I}(i)$. By the hypothesis, for all **U**-expressions which represent

$\prec$-progressive properties of natural numbers, $\prec$-induction for all $j \prec i$ is a rule of **K**$'$, i.e., $(\forall a \epsilon \mathrm{U})[P(a) \rightarrow (\forall \mathrm{j} \epsilon \mathrm{N})[\mathrm{j} \prec \mathrm{i} \rightarrow a\mathrm{j}]]$. Since '$a$' represents a progressive property, this means that 'ai' is in **K**$'$.

4.23. *Lemma.* I(0) is a derived rule of **K**$'$. (The proof, which is an obvious revision of the proof of lemma 4.21, is omitted.)

4.24. It remains to be shown that I(i) is a derived rule of **K**$'$ for all natural numbers i. This will be done in several steps. First, a particular restriction of a property of natural numbers will be defined. Then it will be shown that this restriction I* of I is $\subset$-progressive. Then it will be shown that transfinite $\subset$-induction is a derived rule of **K**$'$. Finally, this result is used to show that transfinite $\prec$-induction is a rule of **K**$'$.

4.25. Let $0(A)$ serve as a metalinguistic abbreviation for the statement that 'A' is a $\subset$-progressive property of natural numbers, i.e., for $(\forall i)\ [(\forall j)\ [j \subset i$ *implies* $A(j)]$ *implies* $A(i)]$.

4.26. *Definition.* The *restriction* A^* of a property A of natural numbers is defined as follows:

$$A^*(i) = \begin{cases} A(0), \text{ if } i = 0; \\ (\forall j)\ [\mathrm{I}(i_j) \textit{ implies } A(i)] \text{ if } i \neq 0. \end{cases}$$

The second part of the definition of A^* asserts that if $\prec$-induction to the indices (4.4) of i is a rule of **K**$'$, then $A^*(i)$ is true.

4.27. *Lemma.* I* is $\subset$-progressive, i.e., 0(I*).

Proof. On the hypothesis

(1) $(\forall j)\ [j \subset i$ *implies* $\mathrm{I}^*(j)]$,

it must be shown that I* (i). By lemma 4.23 and definition 4.26, I* (0). If $i \neq 0$, then by definition 7.26, it must be shown that I(i) is a consequence of

(2) $\mathrm{I}(i_k)$ for all indices of i_k of i.

Since, by lemma 4.22, I is $\prec$-progressive, it is sufficient to show that I(k) is a consequence of

(3) $k \prec i$.

By induction on the rank of k, it will be shown that I(k). If $\rho(k) = 0$, then $k = 0$. I(0) by lemma 4.23. If $\rho(k) > 0$, then $k > 0$. The inductive hypothesis is:

(4) $(\forall n)\ [\rho(n) < \rho(k)$ *and* $n \prec i$ *implies* $\mathrm{I}(n)]$

There are two cases to consider. The first is $i \subset k$. In this case, (3) can be true only by definition 4.7(4), so there is a m such that $k \prec i_m$. Here $\mathrm{I}(k)$ is a consequence of lemma 4.20. The second case is $k \subset i$. By (1), $\mathrm{I}^*(k)$ and

(5) $(\forall m)\ [\mathrm{I}(k_m)$ *implies* $\mathrm{I}(k)]$.

By 4.4, $k_m \prec k$. With (3) we have $k_m \prec i$. Further, $\rho(k_m) < \rho(k)$, so by (4) we have $\mathrm{I}(k_m)$ and by (5) $\mathrm{I}(k)$.

4.28. *Lemma.* For every property A of natural numbers, transfinite $\subset$-induction holds for A^*, i.e., $\mathrm{O}(A^*)$ *implies* $(\forall i)A^*(i)$.
Proof. On the hypothesis $\mathrm{O}(A^*)$, it will be shown that for all i, $A^*(i)$. For $i = 0$, this is a consequence of a minor modification of lemma 4.20. The essential idea of the proof for $i \neq 0$ is this: The first $\ell(i)$ primes are used in the prime decomposition of i. Select an arbitrary integer j (possibly i) which can be decomposed into the first $\ell(i)$ primes. For all $k \subset j$, suppose that $A^*(k)$. On this hypothesis, show $A^*(j)$. By lemma 4.14, $\subset$ is total. The proof summarized above is correct for all natural numbers. Consequently, $(\forall i)A^*(i)$.

Let $\ell(i) = n$ and select arbitrary natural numbers $j_o, j_1, \ldots, j_n$ as the indices of j, i.e., let $j = p_o{}^{j_o} \cdot p_1{}^{j_1} \cdot \ldots \cdot p_n{}^{j_n}$. For all $k \subset j$, suppose $A^*(k)$. By definition 4.7(1) and (3), $k \subset j$ if there is an m such that $k_m \prec j_m$, and for all $q > m$, $k_q = j_q$ or if $k = 0$. For all m such that $0 \leqslant m \leqslant n$,

$$(*)\quad (\forall k_m\ [k_m \prec j_m \textit{ implies } (\forall k_o) \ldots (\forall k_{m-1})\, A^*(p_o{}^{k_o} \cdot p_1{}^{k_1} \cdot \ldots \cdot p_{m-1}{}^{k_{m-1}} \cdot p_m{}^{k_m} \cdot p_{m+1}{}^{j_{m+1}} \cdot \ldots \cdot p_n{}^{j_n})].$$

In order to simplify the following discussion, let '$D(j_m, \ldots, j_n)$' serve as a metalinguistic abbreviation for (*), above, and let '$B(j_m, \ldots, j_n)$' serve as a metalinguistic abbreviation for: $(\forall k_o) \ldots (\forall k_{m-1})A^*(p_o{}^{k_o} \cdot \ldots \cdot p_{m-1}{}^{k_{m-1}} \cdot p_m{}^{j_m} \cdot \ldots \cdot p_n{}^{j_n})$. Of course, these abbreviations apply for $m = 0, 1, \ldots, n$. Note that $B(j_o, \ldots, j_n)$ is $A^*(j)$.

It is straightforward to verify that $D(j_o, \ldots, j_n)$ *and* ... *and* $D(j_n)$ *implies* $B(j_o, \ldots, j_n)$. This is true because A^* is $\subset$-progressive by definition 4.7(2); the antecedent of the implication is equivalent to $(\forall k[k \subset j \textit{ implies } A^*(j)]$.

By induction on m, it will now be shown that for all m such that

$0 \leqslant m \leqslant n$:

(**) $D(j_m, \ldots j_n)$ *and . . . and* $D(j_n)$ *implies* $B(j_o, \ldots, j_n)$.

It has already been shown for $m = 0$. The hypothesis of the induction is that (**) is true for m, and the inductive step is to show that (**) is true for $m + 1$. This is equivalent to giving a proof of $B(j_{m+1}, \ldots j_n)$ on the hypothesis: $D(j_{m+1}, \ldots, j_n)$ *and . . . and* $D(j_n)$ *implies* $[(\forall r)\ [D(r, j_{m+1}, \ldots, j_n)$ *implies* $B(r, j_{m+1}, \ldots, j_n)]]$. It is only necessary to give a proof of $B(j_{m+1}, \ldots, j_n)$ on the hypothesis $(\forall r)\ [D(r, j_{m+1}, \ldots, j_n)$ *implies* $B(r, j_{m+1}, \ldots, j_n)]$. Using the definition of D, this hypothesis is: $(\forall r)\ [(\forall k)\ [k \prec r$ *implies* $B(k, j_{m+1}, \ldots, j_n)]$ *implies* $B(r, j_{m+1}, \ldots, j_n)]$, which asserts that B is $\prec$-progressive in its first argument. If $\prec$-induction to some number s_m is available, we can conclude $B(s_m, j_{m+1}, \ldots, j_n)$. The definition of A^* which is used in the definition of B is such that $\prec$-induction to any index of j holds. Therefore, $(\forall s_m) B(s_m, j_{m+1}, \ldots, j_n)$. This was to be shown to complete the proof of (**).

Therefore, we have $D(j_n)$ implies $B(j_n)$, that is, $(\forall r_n)\ [r_n \prec j_n$ *implies* $B(r_n)]$ *implies* $B(j_n)]$ for all j_n. This is the $\prec$-progressiveness of B. Again, since an essential part of the definition of A^* is that $\prec$-induction holds to any index j_n of j, we have $(\forall s) B(s)$. Therefore, we have $A^*(i)$. That is, the $\subset$-progressive property A^* applies to all natural numbers. This completes the proof of the lemma.

4.29. *Theorem.* $\prec$-induction to any natural number is a derived rule of **K**$'$, i.e., $(\forall i)\ \mathrm{I}(i)$.

Proof. By lemma 4.26, I^* is $\subset$-progressive. By lemma 4.28, $(\forall i)\mathrm{I}^*(i)$. It remains to be shown that I itself is a property of all natural numbers. For all $i \neq 0$, if $\mathrm{I}(i_j)$ for all indices i_j of i, then $\mathrm{I}(i)$. By induction on the rank of i, it will now be shown that $(\forall i)\mathrm{I}(i)$. In case $\rho(i) = 0$, then $i = 0$ and $\mathrm{I}(0)$ by lemma 4.22. In case $\rho(i) > 0$ and $(\forall k)\ [\rho(k) < \rho(i)$ *implies* $\mathrm{I}(k)]$; since $\rho(i_j) < \rho(i)$, $\mathrm{I}(i_j)$ for all indices of i, therefore $\mathrm{I}(i)$.

4.30. This completes the proof of the statement that transfinite induction to constructible ordinals is a derived rule of **K**$'$. With this result, it is possible to develop the arithmetic of ordinals in **K**$'$ and to prove in **K**$'$, for example, that 5 is the first fixed point of the function f defined by the identity $f(i) = exp_o(i,3)$. This is straightforward since the proofs in section 13 of Chapter 4 of

Schütte [22] can be completed in **K**′ using the derived rule of $\prec$-induction (theorem 4.29).

5. Summary

5.1. This essay presents two proofs of the statement that the theory of constructible ordinals can be developed in **K**′. Furthermore, it has been shown that transfinite induction to constructible ordinals is a derived rule of **K**′. This result was established using only induction on the natural numbers (in their usual order).

5.2. In section 3, it was shown that the Church-Kleene ordinals are represented in **K**′ and that the set of these ordinals is completely represented in **K**′. In addition, the arithmetic of constructible ordinals was developed in **K**′. This development includes the fixed points of monotone increasing continuous functions of ordinals.

5.3. In section 4, it was shown that Schütte's ordinals can be dealt with in **K**′. These ordinals are obtained by defining another order relation on the natural numbers. This order relation was shown to be completely represented in **K**′. Then, it was shown that transfinite induction on these ordinals is a derived rule of **K**′. This result suggests a proof of the statement that the addition of a consistent form of material implication would provide a proper extension of **K**′.

5.4. The derived rule of induction may be stated using the conventions of 4.19 as follows: $(\forall a \epsilon \mathrm{U})\ [(\forall \mathrm{j} \epsilon \mathrm{N})\quad (\forall \mathrm{j} \epsilon \mathrm{N})\ [\mathrm{j} \prec \mathrm{i} \rightarrow a\mathrm{j}] \rightarrow a\mathrm{i}]$ *implies* $(\forall \mathrm{k} \epsilon \mathrm{N}) a\mathrm{k}$ *is in* **K**′]. If a rule for material implication sufficiently restrictive to avoid the Curry paradox were available in **K**′, the proof given in section 4 could essentially be carried out inside **K**′ to give a proof of the formula $(\forall a)\ [(\forall \mathrm{i})\ [\mathrm{Ni} \supset [(\forall \mathrm{j})\ [\mathrm{Nj} \supset [\mathrm{j} \prec \mathrm{i} \supset a\,\mathrm{j}]\,] \supset a\mathrm{i}]\,] \supset (\forall \mathrm{k})\ [\mathrm{Nk} \supset a\mathrm{k}]\,]$.Thus, with the addition of implication, this formula would be a theorem of the system. It appears that it cannot be formulated in terms of the primitives of **K**′. This question will be considered in a future paper.

5.5. The system K′ is clearly sufficient to deal with the metamathematical operations which Schütte used to develop his proof theory. For example, the assignment of Gödel numbers to formulas, the assignment of ordinals to formulas, etc. can be completed with K′. Using the derived rule of transfinite induction, it appears to be possible to do all of Schütte's proof theory within K′. Thus K′ provides a consistent metalanguage for this proof theory.

Acknowledgment

It is a pleasure to acknowledge the contributions of Professor William B. Easton of Rutgers University to this essay. Conversations with Professor Easton motivated this study, and he provided helpful advice while the work was in progress. The errors are, of course, the responsibility of the author.

References

[1] Church, A. 1938. The constructive second number class, *Bulletin of the American Mathematical Society* 44: 224-232.
[2] ——, 1942. *The calculi of lambda-conversion.* Princeton: Princeton University Press.
[3] Church, A., and Kleene, S. C. 1937. Formal definitions in the theory of ordinal numbers. *Fundamenta Mathematicae* 28: 11-21.
[4] Feys, R. and Fitch, F. B. 1968. *Dictionary of symbols of mathematical logic.* Amsterdam: North Holland.
[5] Fitch, F. B. 1942. A basic logic. *Journal of symbolic logic* 7: 105-14.
[6] ——. 1944. Representations of calculi. Ibid. 9: 57-62.
[7] ——. 1948. An extension of basic logic. Ibid. 13: 95-106.
[8] ——. 1953. A simplification of basic logic. Ibid. 18: 317-25.
[9] ——. 1953. Self-referential relations. *Proceedings of the 11th International Congress of Philosophy, Brussels, August 20-26* 14: 121-27.
[10] ——. 1954. A definition of negation in extended basic logic. *Journal of symbolic logic* 19: 29-36.
[11] ——. 1956. Recursive functions in basic logic. Ibid. 21: 337-46.
[12] ——. 1956. A definition of existence in terms of abstraction and disjunction. Ibid. 22: 343-44.
[13] Gödel, K. 1967. On formally undecidable propositions of principia mathematica and related systems, I. In *From Frege to Gödel,* J. van Heijenoort, ed. Cambridge, Massachusetts: Harvard University Press.
[14] Guard, J. G. 1964. Lecture notes on recursive functions and mathematical machines. Department of Mathematics, Princeton University.
[15] Hermes, H. 1965. *Enumerability, decidability, computability*, pp. 34-36. Berlin: Springer Verlag.
[16] Lorenzen, P. 1955. *Einfuhrüng in die operativ mathematik.* Berlin: Springer Verlag.
[17] Lorenzen, P., and Myhill, J. 1959. Constructive definition of certain analytic sets of numbers. *Journal of symbolic logic* 24: 37-49.
[18] Myhill, J. 1952. A finitary metalanguage for extended basic logic. Ibid. 17: 164-78.
[19] Orgass, R. J. 1967. A mathematical theory of computing machine structure and programming. (Ph.D. diss., Yale University. Reprinted as IBM Research Paper RC-1863, July 1967.

[20] ——. 1968. Identity of **R** in **K**′. IBM Research Paper RC-2186, August 1968.
[21] —— and Fitch, F. B. 1969. A theory of computing machines. *Studium Generale* 22: 83–104.
[22] Schütte, K. 1960. *Beweistheorie.* Berlin: Springer Verlag.

8: HASKELL B. CURRY

Representation of Markov Algorithms by Combinators

1. Introduction

It is well known that every partial recursive numerical function can be represented by a combinator.[1] By Church's thesis this implies that any effective process can be so represented via the detour through Gödel enumeration. There is some interest, however, in a direct representation, without going through this detour, of certain effective processes, such as Turing machines and Markov algorithms, which do not explicitly refer to numbers. At first glance this seems to require operators, called discriminators,[2] which cannot be adjoined to combinatory logic without producing inconsistency.[3] To avoid these contradictions it is necessary to weaken the system. Thus Kearns [5] showed that if the rule (μ) is dropped, such discriminators can be adjoined consistently, and the resulting system is adequate to represent an arbitrary Turing machine. More recently, Shabunin [9] proposed a system with a weakened rule (μ) which can represent an arbitrary Markov algorithm.

The aim of this paper is to show that such restrictions on (μ) are unnecessary. The production of contradictions depends on two assumptions. The first of these is that the discriminators be total functions, giving a decision for any obs whatever. The second is that a word formed as a sequence of the letters $a_0, a_1, \ldots, a_n$ in the alphabet of the algorithm (or Turing machine) must be represented as

(1) $\quad a_0 a_1 a_2 \ldots a_{n-1}$

where juxtaposition indicates application. According to the usual interpretation of application, the expression (1) means the application of a_0 as function to a_1 as argument to form a new function which is in turn applied to a_2, and so on. From the standpoint of this interpretation, this way of representing a word seems rather unnatural. On the other hand, if we represent a word as a finite sequence or n-tuple, from which the constituents can be recovered by ordinary[4] combinators, and if we admit that discriminators can be partial functions, then at least some of the cited processes can be represented rather simply in terms of combinators, together with a certain basic discriminator. This paper will show this in detail for a Markov algorithm. The consistency of the resulting system follows by the general theorem (due to Hindley) of [3, §12A4].[5]

The notation of this paper will be that of [2,3][6] except that special brackets will be omitted around symbols for numerals (including variables taking numerical values) and numerical functions, it being understood that these symbols denote obs when they appear in formulas in places suitable for names of obs, and otherwise (e.g. as subscripts) have their usual meaning. Other deviations and additions will be noted as they occur.

In the sequel §2 will be devoted to the techniques of finite sequences as such, and §3 to some explanations concerning Markov algorithms. The details of the representation will then follow in §4, and some concluding remarks in §5.

2. *Finite Sequences*

A technique for dealing with finite sequences was developed in [3, §13C]. There, after an extended discussion of various ways of representing n-tuples, we emerged with a sequence of combinators D_n, such that for any $X_1, \ldots, X_n$, a special atom #,[7] and any numeral $k \geqslant 0$,[8]

$$(2) \quad \mathsf{D}_{n+1} X_1 \ldots X_n \# k \geqslant \begin{cases} X_{k+1} & \text{if } k < n, \\ \# & \text{if } k \geqslant n, \end{cases}$$

and there is a D_+ such that[9] $\mathsf{D}_n = \mathsf{D}_+ (n-1)$. Here I shall define

$$(3) \quad \langle X_1, X_2, \ldots, X_n \rangle \equiv \mathsf{D}_{n+1} X_1 \ldots X_n \#,$$

so that we have

$$(4) \quad \langle X_1, X_2, \ldots, X_n \rangle (k-1) = \begin{cases} X_k & \text{if } 1 \leqslant k \leqslant n, \\ \# & \text{if } k > n. \end{cases}$$

In the case where the $X_1, \ldots, X_n$ are letters $a_1 \ldots, a_n$ in an alphabet L, # is not a letter of L, and there is a discriminator which separates L from #, it was shown in [3, §15C4–5] that there exist combinators ⟦length⟧ and ⟦cat⟧[10] such that

(5) $[\![\text{length}]\!] \langle a_1, a_2, \ldots, a_n \rangle = n,$

(6) $[\![\text{cat}]\!] \langle a_1, a_2, \ldots, a_m \rangle \langle b_1 b_2, \ldots, b_n \rangle =$
$\langle a_1, a_2, \ldots, a_m, b_1, \ldots, b_n \rangle.$

We postulate here a *basic discriminator* ⟦Id⟧ such that, if a and b are any two letters of $L' \equiv L \cup \{\#\}$, then

$$(7) \qquad [\![\text{Id}]\!]ab = \begin{cases} 0 & \text{if } a \equiv b, \\ 1 & \text{if } a \not\equiv b. \end{cases}$$

A *discriminator* can then be defined as a constant combination that involves[11] ⟦Id⟧. The term 'combinator' will be extended to include discriminators; combinators not involving ⟦Id⟧ will be called *ordinary combinators.* Then the discriminator $[\![1\text{tr}]\!] \equiv \lambda x \cdot 1 \dot{-} [\![\text{Id}]\!]x\#$ will separate # from L. Such an ⟦Id⟧ can be defined by a finite table, thus:

$$[\![\text{Id}]\!]a_i a_i = 0, \ [\![\text{Id}]\!]\#\# = 0,$$

$$[\![\text{Id}]\!]a_i a_j = 1 \ \text{ if } \ i \neq j, \ [\![\text{Id}]\!]a_i\# = [\![\text{Id}]\!]\#a_i = 1,$$

where a_i, a_j are in L. Thus, as noted in §1, the adjunction of ⟦Id⟧ does not lead to inconsistency.

We now program some particular operations which will be useful later. The first of these is a discriminator ⟦Wid⟧ such that for any L-words P and Q we have

$$(8) \qquad [\![\text{Wid}]\!]PQ = \begin{cases} 0 & \text{if } P = Q, \\ 1 & \text{if } P \neq Q. \end{cases}$$

To do this, observe how we can test the equality of P and Q. We begin by testing $P0 = Q0$. Suppose that for some j we are testing $Pj = Qj$. If the test is negative, then $P \neq Q$; if the test is positive, we proceed to test $P(j + 1) = Q(j + 1)$, provided we have not reached the end of P. If we have reached the end of P, then we are through, and $P = Q$; for if Q were shorter than P we should have had a negative test at an earlier stage, and if Q were longer than P we should have a negative test for j.

All this can be programmed by the combinator ⟦Gp⟧ of [3, §13A2, p. 221], which has the property that for all X, Y, Z, U

(9) $$[\![\mathrm{Gp}]\!]XYZU = \begin{cases} XU & \text{if } ZU = 0, \\ Y([\![\mathrm{Gp}]\!]XYZ)\,U & \text{if } ZU = m+1 \text{ for some } m. \end{cases}$$

Here put

$$Z \equiv \lambda z \cdot 1 \dot{-} [\![\mathrm{Id}]\!]\,(xz)\,(yz),$$

$$X \equiv \lambda z \cdot 1\ (= \mathsf{K}1), \text{ and}$$

$$Y \equiv \lambda uz \cdot \langle 0, u(\sigma z)\rangle\,([\![\mathrm{Id}]\!]\,\#(xz)).$$

Note that X, Y, and Z contain x and y free; P is to be substituted for x and Q for y. Then

(10) $$[\![\mathrm{Wid}]\!] \equiv \lambda xy \cdot [\![\mathrm{Gp}]\!]XYZ0$$

will be the required discriminator.

In terms of $[\![\mathrm{Wid}]\!]$, we can program a termination for a sequence of words which till tell us when we have reached the end of the sequence. For, let P be a sequence of words $X_1, \ldots, X_n$. Let ω be a letter of L not occurring in any of the $X_1, \ldots, X_n$ (if necessary extending L). Then $P' \equiv \langle X_1, \ldots, X_n, \langle\omega\rangle\rangle$ will be such that $P'(k-1) = X_k$ for $1 \leqslant k < n$, $P'n = \langle\omega\rangle$. In terms of this notion we can define length, concatenation, etc. for sequences of words. But we shall not need this extension here.

Remark. If instead of $[\![\mathrm{Id}]\!]$ we use an extension $[\![\mathrm{Id}]\!]^*$, defined by the following finite table, in which a is a fixed letter of L, b any letter of L, if there is one, distinct from a,

	a	b	#	#0	#a
a	0	$[\![\mathrm{Id}]\!]ab$	1	1	1
b	$[\![\mathrm{Id}]\!]ab$	0	1	1	1
#	1	1	0	0	1
#0	1	1	0	0	1
#a	1	1	1	1	0

then we can program a termination signal as follows: Let $C_1 \equiv \lambda x \cdot [\![\mathrm{Id}]\!]^*\,\#(x0)$, $C_2 \equiv \lambda x \cdot [\![\mathrm{Id}]\!]^*\,\#(xa)$. Then

$$C_1X = \begin{cases} 1 & \text{if } X \text{ is a nonempty word,} \\ 0 & \text{if } X \equiv \# \text{ or } \langle\,\rangle \text{ (which is } \mathsf{K}\#), \end{cases}$$

$$C_2X = \begin{cases} 0 & \text{if } X \equiv \mathsf{K}\#, \\ 1 & \text{if } X \equiv \#. \end{cases}$$

Let $C_3 \equiv \lambda x \cdot \langle 1, 1 \dot{-} C_2x\rangle\, (1 \dot{-} (C_1 x))$, then $C_3\# = 0$; $C_3X = 1$ if X is a word, empty or nonempty. Thus C_3 may be used to detect the end of a sequence of words, just as $[\![\mathrm{Id}]\!]$ detects the end of a sequence of letters. However, this process involves some complexity, and the above device is simpler, especially since the stop sign of the algorithm is available as a suitable ω.

In a similar way, we can construct a discriminator for testing for an occurrence of a fixed letter, say a, i.e., if W is an L-word and a is an L-letter,

$$(11) \qquad [\![\mathrm{Oc}]\!]aW = \begin{cases} 0 & \text{if there is an occurrence of } a \text{ in } W \\ 1 & \text{if there is no occurrence of } a \text{ in } W. \end{cases}$$

This is given by $[\![\mathrm{Oc}]\!] = \lambda xw \cdot [\![\mathrm{Gp}]\!]\, XYZ0$, where

$$Z \equiv \lambda z \cdot [\![\mathrm{Id}]\!]\, x(wz),$$

$$X \equiv \mathsf{K}0,$$

$$Y \equiv \lambda yz \cdot \langle 1, y(\sigma z)\rangle\, ([\![\mathrm{Id}]\!]\#(wz)).$$

Slightly more complex is a discriminator $[\![\mathrm{Er}]\!]$ such that $[\![\mathrm{Er}]\!]a$ erases all occurrences of a in W. To define this, let

$$W_0 \equiv \langle\,\rangle,$$

$$W_{j+1} \equiv [\![\mathrm{cat}]\!]\, W_j\, (\langle\langle\,\rangle, \langle Wj\rangle\rangle\, ([\![\mathrm{Id}]\!]a(Wj))).$$

Let

$$J \equiv \lambda xwzt \cdot [\![\mathrm{cat}]\!]\, t\, (\langle\langle\,\rangle, \langle wz\rangle\rangle\, ([\![\mathrm{Id}]\!]x(wz))).$$

Then

$W_{j+1} = JaWj(W_j)$. Hence $W_j = \mathsf{R}\,\langle\,\rangle\,(JaW)\,j$. Abstracting with respect to a and W, we have

$$(12) \qquad [\![\mathrm{Er}]\!] \equiv \lambda xw \cdot \mathsf{R}\,\langle\,\rangle\,(Jxw)\,([\![\mathrm{length}]\!]w).$$

Note that if there is no occurrence of a in W, then $[\![\mathrm{Er}]\!]aW = W$.

Next we consider a discriminator $[\![\mathrm{Seg}]\!]$ which extracts from a word W the sequence of length k beginning with the $(h + 1)$st constituent of W, i.e., if $W \equiv \langle a_1, a_2, \ldots, a_n\rangle$, then

(13) $[\![\text{Seg}]\!]hkW = \langle a_{h+1}, \ldots, a_{h+k}\rangle = \langle Wh, \ldots, W(h+k-1)\rangle$

Here let $T_0 \equiv \langle\,\rangle$, $T_{j+1} \equiv [\![\text{cat}]\!]T_j\langle W(h+j)\rangle$. Let

$$J \equiv \lambda xwzt \cdot [\![\text{cat}]\!]\, t\langle w(x+z)\rangle.$$

Then $T_{j+1} = JhWj(T_j)$. Hence $T_j = \mathsf{R}\langle\,\rangle(JhW)j$, and the desired sequence is T_k. Abstracting with respect to h,k,W, we have

(14) $[\![\text{Seg}]\!] \equiv \lambda xyw \cdot \mathsf{R}\langle\,\rangle(Jxw)y.$

This of course will give nonsensical results if $h + k > [\![\text{length}]\!]\, W$. We shall apply it only under the restriction $h + k \leqslant n$.

Finally, the result of placing the indicated sequence by a word V is $[\![\text{Rp}]\!]\, hkWV$, where

$$[\![\text{Rp}]\!] \equiv \lambda xywv \cdot [\![\text{cat}]\!]\,([\![\text{Seg}]\!]0xw)\,([\![\text{cat}]\!]\, v\,([\![\text{Seg}]\!]\,(x+y)\,(n \dot{-} (x+y))w)\,).$$

3. Markov Algorithms

No attempt will be made here to describe a Markov algorithm fully.[12] I shall simply state some conventions and special assumptions needed for the present treatment.

We consider an algorithm $\mathfrak{A}$ which operates on words W in an alphabet L_0. The scheme of the algorithm is an ordered set of commands

(15) $U_i \rightarrow V_i$,

with or without a stop signal, usually in the form of a dot after the '→'. Here the letter ω will be used for the stop signal; when present it will be regarded as the first letter of V_i. The words U_i and V_i will usually contain letters, called *auxiliary letters*, not in L_0; the alphabet formed by adjoining these letters, except ω, to L_0, will be called L_1. (Then, in the usual terminology, $\mathfrak{A}$ is said to be over L_0 and in L_1.) Let L_2 be the alphabet formed by adjoining ω to L_1. It will be supposed that # is not a letter of L_2, and the alphabet formed by adjoining # to L_2 will be called L_3. Then the discriminator $[\![\text{Id}]\!]$ is to be defined over L_3. Thus # is not a letter of any alphabet used in the algorithm itself, but is an extraneous letter useful in the combinatory representation.

In many algorithms it is necessary to start by inserting an auxiliary letter, here called the *starter*, at the beginning of W. Here the starter will be designated α. The usual way of doing this is to have a command

$$(16) \quad \rightarrow \alpha$$

at the end of the scheme (15), all U_i in previous commands having at least one auxiliary letter. Then the algorithm should be so designed that once started it is not blocked, i.e., it reaches a stop signal and stops before coming to (16) again. However, in the combinatory representation it is just as easy to program a preliminary transformation for inserting the starter before the algorithm proper begins. Since the preliminary transformation is not activated but once, it is not necessary to avoid blocking. Thus we are dealing with a slight generalization of a Markov algorithm; it may be called a *Markov algorithm with initial starter.*

Although blocking can always be avoided by using the closing order[13]

$$(17) \quad \rightarrow \omega$$

at the end of (15), yet to obtain complete generality we must allow for its occurrence in $\mathfrak{A}$. We can do this by providing an end signal for the termination of the sequence $U_1, U_2, \ldots, U_p$ as described in §2.

4. Representation of the Algorithm

Given the algorithm $\mathfrak{A}$, with starter α and stop signal ω, let

$$(18) \quad \begin{aligned} U &\equiv \langle U_1, U_2, \ldots, U_p, \langle\omega\rangle\rangle, \\ V &\equiv \langle V_1, V_2, \ldots, V_p\rangle. \end{aligned}$$

These are constants determined only by $\mathfrak{A}$. Let W_0 be the word being operated on. Further let (the subscript i will be omitted when clear from the context)

$$(19) \quad \begin{aligned} k_i &\equiv [\![\text{length}]\!]\, U_i, \\ m_i &\equiv [\![\text{length}]\!]\, V_i, \\ n &\equiv [\![\text{length}]\!]\, W. \end{aligned}$$

We seek a combinator G_0 such that

(20) $G_0 UVW_0 \geqslant W' \rightleftarrows \mathfrak{A}(W_0) = W'$.

In the following explanation combinators are defined as functional abstracts with bound variables $x_1, x_2, x_3, x_4, x_5, y_1, y_2, y_3$. In application, these variables take values as follows: for x_1, U; for x_2, V; for x_3, W_0 and the words it is changed into; for x_4, $i-1$ where U_i is the left side of the command being considered; for x_5 the beginning of the segment being tested for an occurrence of U_i; and y_1, y_2, y_3 for the function to be iterated in the various cases of ⟦Gp⟧ (corresponding to the appearance of ⟦Gp⟧XYZ on the right side of (9)). In the informal discussion it is convenient to use these letters as standing for the corresponding values.

There are four principal combinators, G_0, G_1, G_2, G_3, of which the last three are headed by ⟦Gp⟧. The actions of these are as follows: G_0 inserts the starter and passes to G_1; if there is no starter, $G_0 \equiv G_1$. G_1 tests for the presence of ω in x_3; if the test is negative, G_1 sets x_4 at 0 and passes to G_2; if the test is positive, it erases ω and stops. G_2 tests x_3 for an occurrence of $x_1 x_4$, using G_3; if the test is positive, G_2 performs the substitution and passes to G_1; if the test is negative, then G_2 repeats with $x_4 + 1$ in the place of x_4, provided that $x_1(x_4 + 1)$ is not the terminus of U, in which case it stops. G_3 tests x_5 for an occurrence of $x_1 x_4$ beginning there; if x_5 is so large that the rest of x_3 is too short, then it sends a signal 1 to G_2; if the test is positive a signal 0 is sent to G_2; if the test is negative and x_5 is not too large, then G_3 increases x_5 by 1 and repeats.

We now proceed to formal definitions of these combinators. First we have

$$
\begin{aligned}
G_0 &\equiv \lambda x_1 x_2 x_3 \cdot G_1 x_1 x_2 ([\![\text{cat}]\!] \langle\alpha\rangle x_3), \\
G_1 &\equiv \lambda x_1 x_2 \cdot [\![\text{Gp}]\!] X_1 Y_1 Z_1, \\
G_2 &\equiv \lambda x_1 x_2 x_3 \cdot [\![\text{Gp}]\!] X_2 Y_2 Z_2, \\
G_3 &\equiv \lambda x_1 x_2 x_3 x_4 \cdot [\![\text{Gp}]\!] X_3 Y_3 Z_3,
\end{aligned} \tag{21}
$$

where X_j, Y_j, Z_j are obs, possibily containing some or all of the variables bound in the prefix to G_j, which we shall now determine. For $j = 1$ we have

$$Z_1 \equiv \lambda x_3 \cdot [\![\text{Oc}]\!]\, \omega x_3,$$

(22) $X_1 \equiv \lambda x_3 \cdot [\![\mathrm{Er}]\!]\, \omega x_3,$

$Y_1 \equiv \lambda y_1 x_3 \cdot G_2 x_1 x_2 x_3 0 y_1.$

We now turn to the determination of Y_2 and Z_2. We leave X_2 until after we have finished with G_3.

(23)
$$Z_2 \equiv \lambda x_4 \cdot G_3 x_1 x_2 x_3 x_4 0,$$
$$Y_2 \equiv \lambda y_2 x_4 y_1 \cdot \langle x_3, y_2(\sigma x_4) y_1\rangle\, ([\![\mathrm{Wid}]\!]\, \langle\omega\rangle\, (x_1(\sigma x_4))).$$

Next we turn to G_3. The Z_3 will give a zero output as soon as there is a decision as to whether there is an occurrence of $x_1 x_4$ in x_3. The decision will be negative if $x_5 + k > n$; it will be positive if $x_5 \leqslant n \dot{-} k$ and we have $H_3 x_5 = 0$, where

(24) $H_3 \equiv \lambda x_5 \cdot [\![\mathrm{Wid}]\!]\, (x_1 x_4)\, ([\![\mathrm{Seg}]\!] x_5 k x_3).$

Otherwise there is no decision, and we must increase x_5. Thus we set[14]

(25)
$$Z_3 \equiv \lambda x_5 \cdot \langle H_3 x_5, 0\rangle\, |x_5 + k \dot{-} n|,$$
$$X_3 \equiv \lambda x_5 \cdot |x_5 + k \dot{-} n|,$$
$$Y_3 \equiv \lambda y_3 x_5 \cdot y_3(\sigma x_5).$$

We now turn to X_2. Note that we come out to X_2 only if there is an $x_5 \leqslant n \dot{-} k$ such that $H_3 x_5 = 0$. Let h be the least such x_5, i.e.,

(26) $h = [\![\mathrm{Pe}]\!] H_3 0.$

Then let

(27) $H_2 \equiv \lambda x_4 \cdot [\![\mathrm{Rp}]\!] h k x_3 (x_2 x_4).$

(Note that h and k depend on x_3 and x_4.) Then we can set

(29) $X_2 \equiv \lambda x_4 y_1 \cdot y_1(H_2 x_4).$

The proof will now be given that the G_0 defined by (21) to (29) satisfies (20). In doing this I shall suppose first that the reductions in question are $\lambda\beta\delta$ reductions.[15] First suppose the right side of (20) holds. Then there is a reduction $\mathfrak{D}_1$, which carries out exactly the instructions of $\mathfrak{A}$. To characterize $\mathfrak{D}_1$ more precisely, we first note that (9) does not hold in the sense of a reduction from left to right. Rather, if ⟦Gp⟧ is defined as in [3, §13A(30)], then in the sense of (incomplete) normal $\lambda\beta\delta$ reduction, we have

$$[\![\mathrm{Gp}]\!]\, XYZ \geqslant N \equiv \lambda t \cdot QXYZ\,(Zt)\,(QXYZ)\,t,$$

where Q is defined as in [3], and for any T

$$NT \geqslant \begin{cases} XT & \text{if } ZT = 0 \\ YNT & \text{if } ZT \geqslant 1, \end{cases}$$

so that NT is a stage in the reduction of ⟦Gp⟧ $XYZT$, and if either NT or ⟦Gp⟧$XYZT$ has a normal form, so does the other. Then $\mathfrak{D}_1$ is a normal $\lambda\beta\delta$ reduction, except that when an N_j comes to a senior position, i.e., a position such that it would be at the beginning of the next redex to be contracted, in the present case head position, it is frozen and not thawed out until the reduction (with the limitation mentioned) has reached normal form, at which time it is reactivated, and, in effect, iterates G_j. Then $\mathfrak{D}_1$ reduces G_0UVW_0 to W' if any only if the right side of (20) holds. Thus (20) holds in the sense from right to left.

Conversely, suppose there is a reduction $\mathfrak{D}_2$, not necessarily the same as $\mathfrak{D}_1$, from G_0UVW_0 to W'. By the standardization theorem,[16] we can suppose, since W' is in normal form, that $\mathfrak{D}_2$ is a normal reduction, and hence it differs from $\mathfrak{D}_1$ only in that the N_j are not frozen, but are activated immediately when they reach senior position. Thus certain internal contractions allowed in $\mathfrak{D}_1$ are not made at that point. But it can be shown[17] that these contractions have to be made later in $\mathfrak{D}_2$, so that if $\mathfrak{D}_2$ reaches a normal form W', $\mathfrak{D}_1$ does also. Thus, by the end of the preceding paragraph, we have (20) in the sense from left to right.

This completes the proof. Note that it can further be shown that neither $\mathfrak{D}_1$ nor $\mathfrak{D}_2$ requires the use of (ξ) (contraction within the scope of a λ-prefix), so that they are easily transformed into weak reductions.

5. Conclusion

It has thus been shown that we do not have to weaken the rule (μ) or other accepted principles of combinatory logic in order to represent a Markov algorithm. Very likely an analagous statement is true for Turing machines or other forms of effective process. Of course, it is remarkable that the weakened systems of Kearns and Shabunin will do as much as they do. But there is a certain artificiality about those systems, not only for the reason stated in §1, but because (μ) is a natural property of equality, and of reduction conceived as generated by replacements. We have seen that this artificiality is not necessary; and the claim that these systems are actually more powerful than combinatory logic can hardly be maintained.

References

[1] Curry, Haskell B. 1963. *Foundations of mathematical logic.* New York: McGraw-Hill.

[2] Curry, Haskell B., and Feys, Robert. 1958. *Combinatory logic, vol. I.* Amsterdam: North-Holland.

[3] Curry, Haskell B.; Hindley, J. Roger; and Seldin, Jonathan P. 1972. *Combinatory logic, vol. II.* Amsterdam: North-Holland.

[4] Hindley, J. Roger; Lercher, Bruce; and Seldin, Jonathan P. 1972. *Introduction to combinatory logic.* Cambridge: Cambridge University Press.

[5] Kearns, John T. 1969. Combinatory logic with discriminators. *J. Symbolic Logic* 34:561–75.

[6] Kleene, Stephen C. 1936. λ-definability and recursiveness. *Duke Mathematical Journal* 2:340–53.

[7] Markov, A. A. 1954. *Teoriya algorifmov* (Theory of algorithms). Trudy M. Inst. im Steklov, no. 42. Moscow, 1954. Translated by J. J. Schorr-Kon; Jerusalem: S. Monson, 1961.

[8] Mendelson, Eliott. 1964. *Introduction to mathematical logic.* Princeton, N.J.: Van Nostrand.

[9] Shabunin, L. V. 1972. Variant kombinatornoi logiki s diskriminatorami (A variant of combinatory logic with discriminators). Vestnik Moskov. Univ. Ser. I Mat. Meh. 27: 53–58.

[10] Venturini-Zilli, Marisa. 1965. λ-K formulae for vector operators. Bulletin of the International Computation Center, Rome. 4:157–74.

PART III IMPLICATION AND CONSISTENCY

9: ALAN ROSS ANDERSON

Fitch on Consistency

Frederic Fitch's search for a "powerful" and "demonstrably consistent" system of mathematical logic is well known to all of us from his 1938 paper (references are to the bibliography at the end of this paper) and from his investigation of basic logic (see [10]). "Powerful," in this connection, means having the capacity for developing theories of the natural numbers, the integers, the rationals, the reals, etc., in such a way as to be able to prove the recognizably most important theorems of each of these number systems—a property which is difficult to make precise, but which is still natural and to a certain extent comprehensible. I will suppose for the purpose of this paper that we all have some understanding of this notion, and direct my attention from here on to the meaning of "demonstrably consistent." As a preliminary, however, I will devote a few paragraphs to the historical and metaphysical presuppositions which seem to lie behind his views about how power and consistency are intertwined.

Fitch has, of course, investigated other abstract mathematical and philosophical structures, but most of his investigations into the foundations of mathematics have been conducted in various forms of the theory of combinators embedded in nonclassical propositional and quantificational systems of logic. These studies have been coupled with what he considers an important metaphysical requirement: that we ought to be able, in some sense, to talk about "everything at once." If metaphysics is conceived of in a Whiteheadean sense as describing the universe once and for all in its entirety, then it ought to be able to give an account of such things as real numbers, chairs, love, doves, hawks, strange rumblings in our stomachs, techniques for landing submarines, the U.S. Steel Corporation, the flight of every sparrow, the best of Caruso's renditions of "Vesti la giubba," the extraordinary predominance of beetles in the animal kingdom, etc.

For reasons which must be sought elsewhere in Fitch's writings, he seems to feel that the holistic view he wants to espouse is closely connected with the logical capacity to encompass a truly universal class within a formal system. That is to say, $x = x$ shall, for example, be true of every entity (in some *very strong* sense of *every*) and, therefore, "$\hat{x}[x = x]$" shall denote the class that has absolutely every entity as a member, including, of course, itself. This requirement is connected in turn with absolutely unlimited abstraction principles, in effect the so-called naive class axiom, which Fitch [11] defended as follows:

> In order to avoid the Curry paradox or the Russell paradox, it would seem to be sufficient to restrict the abstraction rules in some suitable way, for example by invoking the simple theory of types, so that such expressions as '$x \in x$' would be rejected as not well formed. But all proposed restrictions on abstraction, including those of type theory and those of set theory, seem unduly arbitrary and, in the case of the simple theory of types and standard set theory, are not even known to guarantee freedom from contradiction. Consequently there is a good deal of theoretical interest in seeing how far one can go without placing any restriction at all on abstraction. If the abstraction rules are to be used as embodying a valid logical principle, namely the principle of abstraction, then it is hard to see how any limitation can be placed on abstraction that is not itself a purely logical limitation. But there do not seem to be such limitations that are both purely logical and sufficiently nonarbitrary. In other words, our logical intuitions do not appear to tell us clearly exactly how much leeway to grant to the principle of abstraction, so that if we accept it at all as a principle of logic, we should apparently accept it without limitation.

The passion for unlimited abstraction led Fitch [9] to a general solution to a problem concerning self-referential relations in his very weak system K of basic logic [7], which is a subsystem of all the other extended forms of basic logic. including those of [8]; the solution therefore holds in these systems as well. As a corollary of the general result, he can define a class F which has itself as its only member; i.e., it satisfies the condition $(x)[[x \in F] \equiv [x = F]]$. It is without doubt difficult for those immersed in, or perhaps immured by, the Russell-type-theoretical or Zermelo-set-theoretical traditions to attach much intuitive sense to this extraordinary class.

As evidence I cite the fact that I once mentioned to a distinguished "working mathematician" that a set satisfying the remarkable condition displayed above could be defined in the theory of combinators. After peering for a minute or so at the piece of paper on which the condition was written, he handed it back to me with the remark, "That's the kind of thing that makes my head hurt"— and then went back to work.

The line of thought quoted from Fitch in the paragraph above originates so far as I know in Russell, who writes:

> Thus, for example, the proposition "x and y are numbers implies $(x + y)^2 = x^2 + 2xy + y^2$" holds equally if for x and y we substitute Socrates and Plato: both hypothesis and consequent, in this case, will be false, but the implication will still be true. Thus in every proposition of pure mathematics, when fully stated, the variables have an absolutely unrestricted field: any conceivable entity may be substituted for any one of our variables without impairing the truth of our proposition [17, p. 7].

This notion had plausibility at that time, especially because the logical circles in which Russell moved were agreed that there was exactly one correct mathematical or logical sense of "if . . . then –," or "implies." Seventy years later this position seems at least debatable and, given recent discussions of restricted quantification and even restricted assertion [3], I don't see why we must look at the matter in Russell's way, though of course we may. But I won't be fussy here about this question. Rather, I shall simply give Fitch the point for the sake of argument and then try to see where it may lead us.

Of course, many of the members of $\hat{x}[x = x]$ can be explained away, in that more complicated events or phenomena can be satisfactorily accounted for with the help of, or "reduced" to as in Fitch [12], ideas or categories thought to be simpler, or somehow more fundamental, and many philosophers have had reductionist tendencies. Russell, for example in [17, p. 5], pronounced that "the fact that all Mathematics is Symbolic Logic is one of the greatest discoveries of our age," though I doubt that many people believe that nowadays or not, at least, without severe reservations or qualifications.

The foregoing brief historical remarks are intended to set the stage for the topic I will concentrate on in this paper, namely, the second of the two expressions with which I began, "demonstrably consistent," in the hope of winding up with some constructive suggestions about how this matter should best be viewed from Fitch's own point of view.

I

The notion of consistency for formal systems of logic has been defined in several ways, among them (a) consistency with respect to a given transformation (typically transformation of A into $\sim A$), (b) absolute consistency (some well-formed formulas are not theorems), (c) consistency in the sense of Post (no propositional variable is a theorem). For discussion of these and related notions see Church [4, §§17, 45, 54, 55]. Other similar notions concern consistency with respect to some semantical structure such as the multi-valued truth table (e.g., Łukasiewicz, Tarski, Sobociński and others), consistency with respect to some more elaborate algebraic or set-theoretical structure, and the like. Generally these more elaborate consistency definitions are more interesting for completeness proofs than for consistency proofs, since most well-known nonclassical systems of propositional, modal, and quantificational logic can be shown to be consistent simply by showing that under some appropriate mapping they are subsystems of the classical systems, which are known to be consistent.

It should be noted that there are exceptions to this generalization. In the consistent sysems of Angell [2] and McCall [14] there is an arrow dubbed "connexive implication" by McCall, with theorems like $A \to B \to . \sim(A \to \sim B)$, which is no tautology if the arrow is read truth functionally. It proves possible even more daftly to add consistently to the Anderson-Belnap system E of entailment (for references see Meyer and Dunn [15] or Anderson and Belnap [1] the formula $A \to B \to . C \to D$, leading to the consequence $A \to A \to . A \to A$, which is a flat truth-functional contradiction if the intensional arrow is misread as truth-functional material "implication." For present purposes, connexive and relevance implication relations have been cited only as counterexamples to the generalization of the last paragraph and will receive no further discussion here. But I will take the opportunity to dub the system E, together with the first of the two formulas cited just above, "E^f", where the f stands for "freak," indicating that we just don't know what to make of it.

Now much of the discussion of these various definitions of consistency has centered, implicitly if not explicitly, on what is the "best" definition of consistency. This depends, of course, in part on the system under consideration. For the classical two-valued truth-value calculus TV, or "propositional calculus" as it is usually, and confusingly, called, the answer seems clear: TV is consistent if there is no formula A such that $\vdash_{TV} A$ and $\vdash_{TV} \sim A$, and for this example the three definitions of consistency (a)–(c) above are equivalent. But what if a system has no negation? Intuitions stemming from those persuaded that the horseshoe means "if . . . then –," together with the tautology $p \mathbin{\&} \sim p \supset q$, bring to mind that a contradiction in TV leads to what Bob Meyer has called a psychotic break. So by a natural and easy transition we are led to say that a system without negation is inconsistent if it is psychotic, i.e., every formula is provable, and such a system is consistent if it is not psychotic (the more common term for "psychotic" is "incoherent"). A coherent system then has some unprovable formulas and is absolutely consistent in the sense (b) mentioned above, and normally is consistent in the sense of Post, (c) above, as well.

But this won't quite do, even for negation-free systems. Kemeny [13] gives a nice example of a system with one unprovable formula, which any sane man would even so regard as psychotic. Paraphrasing his argument, he asks us to consider a system consisting of object-language propositional variables $d, p_1, \ldots, p_i, \ldots$, where formulas are either variables or else built up out of formulas and $\supset$ as usual (a pure implicational calculus suffices for the example), and where modus ponens for $\supset$ and uniform substitution are rules of inference, except that d may not be substituted in the sole axiom of the system, which is p_1. Clearly this system is both absolutely consistent and consistent in the sense of Post, since the propositional variable d is not a theorem. But every other formula is, and Kemeny says (and we agree) that for this admittedly uninteresting system, neither of the latter definitions are appropriate: they are not at all in accord with the intuitive idea of consistency with which we presumably began, given that we wanted the horseshoe to have some nontrivial interpretation.

A more interesting example of an incoherent system, which in fact possesses a kind of negation and has some other apparently sensible properties as well, arises from the propositional system of

Fitch [8, §20.2, R1–R23] by dropping the restrictions on proof construction of §§1–18.6. This is Fitch's familiar "method of subordinate proofs," with positive and negative introduction and elimination rules for identity, for propositional connectives—except that there are no negative rules for the horseshoe, and no rule of negation introduction (*reductio ad absurdum*, see below)—for the ϵ-relation of membership, and for abstraction. I omit his rules for modality as being irrelevant for present purposes. In this system, where the Russell class R may be defined as $\hat{x}{\sim}[x \epsilon x]$, we can indeed prove that $R\epsilon R .\equiv. {\sim}R\epsilon R$ by the usual moves, but since the law $p \vee {\sim}p$ is not provable in general, the standard way of continuing so as to prove both $R\epsilon R$ and ${\sim}R\epsilon R$ is not available. But the system does succumb to a form of paradoxical argument due to Curry [5], as we now see.

The crux of the difference between Russell's paradox and Curry's is that the Russell class $\hat{x}{\sim}[x\epsilon x]$ involves negation essentially, whereas Curry's class Y_q, defined as $\hat{x}[[x\epsilon x] \supset q]$, does not; we simply chose q arbitrarily, thus allowing proof of every formula and hence incoherence. For facilitating comparisons we use the notation of Fitch [11]. He presents the argument as follows:

1	$\quad\vert\ Y_q \epsilon Y_q$	hypothesis
2	$\quad\vert\ Y_q \epsilon \hat{x}[[x\epsilon x] \supset q]$	1, repetition, definition
3	$\quad\vert\ [Y_q \epsilon Y_q] \supset q$	2, abstraction elimination
4	$\quad\vert\ q$	1, 3, implication elimination
5	$[Y_q \epsilon Y_q] \supset q$	1–4, implication introduction
6	$Y_q \epsilon \hat{x}[[x\epsilon x] \supset q]$	5, abstraction introduction
7	$Y_q \epsilon Y_q$	6, repetition, definition
8	q	5, 7, implication elimination.

Now there is something funny about this argument, in the sense that we used the hypothesis of a hypothetical proof as a minor premise of an application of modus ponens in order to establish a categorical conclusion. This oddity was at least regarded by Fitch as the focal point of the difficulty, so some adjustments had to be made.

But before going on to discuss the required adjustments, I would like to point out another anomaly connected with the Curry argument: it proves to be impossible to add to the rules for the propositional connectives, abstraction, and the epsilon (as summarized in §20.2 of Fitch [8]), the plausible rule

> R? Negative implication introduction. $\sim[p \supset q]$ is a d. c. of p and $\sim q$.

This fact was noted independently by H. P. Galliher and J. R. Myhill, as Fitch [8, p. 127n.] points out, and also independently by the present writer at about the same time. Since a proof has never appeared in print, it might be worth pointing out how a version of the argument goes.

Consider the following two proofs, where now we consider $Y_{\sim[p \supset p]}$ rather than Y_q as before and where, for notational convenience, we use **r** to abbreviate $\sim(p \supset p)$:

I	1	$\quad [Y_{\mathbf{r}} \epsilon Y_{\mathbf{r}}] \supset \mathbf{r}$	hypothesis
	2	$\quad Y_{\mathbf{r}} \epsilon \hat{x}[[x \epsilon x] \supset \mathbf{r}]$	1, abstraction introduction
	3	$\quad [Y_{\mathbf{r}} \epsilon Y_{\mathbf{r}}]$	2, repetition, definition
	4	$\quad \mathbf{r}$	1, 3, implication elimination
	5	$[[Y_{\mathbf{r}} \epsilon Y_{\mathbf{r}}] \supset \mathbf{r}] \supset \mathbf{r}$	1–4, implication introduction

And 5, expanding the definition of **r**, comes to 5′ $[[Y_{\sim[p \supset p]} \epsilon Y_{\sim[p \supset p]}] \supset \sim[p \supset p]] \supset \sim[p \supset p]$:

II	1	$\quad [Y_{\mathbf{r}} \epsilon Y_{\mathbf{r}}]$	hypothesis
	2	$\quad Y_{\mathbf{r}} \epsilon \hat{x}[[x \epsilon x] \supset \mathbf{r}]$	1, repetition, definition
	3	$\quad [Y_{\mathbf{r}} \epsilon Y_{\mathbf{r}}] \supset \mathbf{r}$	2, abstraction elimination
	4	$\quad \mathbf{r}$	1, 3, implication elimination
	5	$[Y_{\mathbf{r}} \epsilon Y_{\mathbf{r}}] \supset \mathbf{r}$	1–4, implication introduction
	6	$\quad p$	hypothesis
	7	$\quad p$	6, repetition
	8	$p \supset p$	6–7, implication introduction
	9	$\sim\sim[p \supset p]$ (i.e., $\sim\mathbf{r}$)	8, double negation introduction
	10	$\sim[[[Y_{\mathbf{r}} \epsilon Y_{\mathbf{r}}] \supset \mathbf{r}] \supset \mathbf{r}]$	5, 9, negative implication introduction (?)

And this formula unpacks into

$$10' \sim[[[Y_{\sim[p\supset p]} \epsilon Y_{\sim[p\supset p]}] \supset\sim[p\supset p]] \supset\sim[p\supset p]].$$

Clearly step 10′ of proof **II** explicitly contradicts the step 5′ of proof **I**, so that if the Fitch system of 1952 is to be thought of as the class of its theorems, addition of a rule for negative implication introduction renders that class inconsistent relative to negation. One solution is immediate, of course: simply abandoning the rule in spite of its natural character. That is the course Fitch [8, §20.48] took, adding that "fortunately these negative implication rules are not required for our purposes, and we shall not use them or assume them."

But even abandoning a rule of negative implication introduction still leaves Curry's disastrous paradox to cope with and, as remarked before, Fitch took as the crucial intuitive point the fact that in Curry's argument we were using some formula which had already been assumed earlier as the hypothesis of an hypothetical subproof in a categorical proof as a premise for modus ponens for the ⊃. In order to avoid Curry's argument he formulated two restrictions on proofs. The Simple Restriction reads: "No item of a main proof or of a subordinate proof can have an elimination rule as a reason if some preceding item of that same proof has an introduction rule as a reason" [8, p. 107]. As Fitch remarked later in [11], however, this restriction, though it avoids the Curry paradox in the formulation given above, also rules out other plausible proofs which we want to recognize intuitively as valid, e.g.,

1	$[p\&q\supset r$	hypothesis
2	p	hypothesis
3	q	hypothesis
4	$p\&q$	2, reiteration, 3 conjunction introduction
5	$[p\&q]\supset r$	1, reiteration
6	r	4, 5 implication elimination.

But the Simple Restriction blocks step 6, and hence, contrary to Fitch's wishes, the Law of Exportation is unprovable in the usual way.

So the Simple Restriction, even though it preserves consistency, is evidently too strong, and Fitch then formulates a more subtle Special Restriction, which serves to save us from Curry's paradox, while being more liberal than the Simple Restriction, though at the expense of considerable complication. As preliminary to stating the Special Restriction, we paraphrase the definition of the relation *is a resultant of* in the following four clauses or conditions: (i) If a formula occurs twice in any proof, regardless of subordination relations in the proof, each is a resultant of the other. (ii) If a formula is a direct consequence of one or more preceding items (formulas or subproofs), then that formula is a resultant of each of those preceding items. (iii) A subordinate proof is a resultant of each of its own formulas and subproofs subordinate to it. (iv) The relation *is a resultant of* is transitive. The Special Restriction as originally stated then requires that no formula *p* may occur in a proof in such a way as to be the resultant of a subordinate proof that (1) has *p* as an hypothesis and (2) contains some formula other than *p* as a step. Examination of the Curry argument displayed above reveals that it violates the Special Restriction. Writing "xRy" to mean that step x is a resultant of step y, which may itself be a subproof, we observe that

7 R 6 by (ii),

6 R 5 by (ii), and

5 R (1–4) by (ii), (iii),

so by (iv), we have

7 R (1–4);

but 1 R 7 by (i)

whence 1 R (1–4) by (iv).

So the Curry argument fails at step 7, since step 1 (= step 7) is a resultant of the subproof (1–4), in violation of the Special Restriction.

Now it might be thought that though the Special Restriction rules out Curry's paradox but does not rule out either of the proofs **I** and **II**, both of which satisfy the restriction, and each of which is consistent separately, we might still obtain incoherence by laying **I** and **II** end-to-end and invoking the rule of negation elimination ("*q* is a direct consequence of *p* and ~*p*") to get an arbitrary formula q as a theorem. If the system is to be identified with the class of its theorems, as is usually done, this would be the expected course, but as Fitch points out in [8, §20.42], there is no reason to believe that the result of laying end-to-end two proofs, both of which satisfy the Special Restriction, will itself result in a sequence of steps satisfying that restriction. And indeed the proofs **I** and **II** above provide a case in point: laying them end to end in either order provides a sequence which fails to satisfy the Special Restriction. For:

I3 R **I**2	by (ii),
I2 R **I**1	by (ii),
I1 R **II**5	by (i), and
II5 R **II**(1–4)	by (ii), (iii),

so, by (iv), we have

I3 R **II**(1–4);

but	**II**1 R **I**3	by (i)
whence	**II**1 R **II**(1–4)	by (iv),

in violation of the Special Restriction, no matter which of **I** and **II** precedes the other in the course of laying then end-to-end. (We note in passing that the Special Restriction may be stated more simply as requiring that no hypothesis p may occur as the resultant of a subproof of which p is the hypothesis, as is shown by the examples above and easily proved in general.)

Before trying to draw some morals from this discussion of the Special Restriction, with which Fitch himself expressed long-felt dissatisfaction in [11], we consider a proposal of Prawitz [16], which we shall call the "Prawitz Restriction." We quote Fitch's formulation of it in [11] as follows:

> No items of a proof or subproof may serve both as direct consequences with respect to an introduction rule and as the major premisses of an elimination rule. (The premisses of all elimination rules will be said to be major premisses except that in the case of implication elimination we call 'p' the minor premiss and '$p \supset q$' the major premiss.)

Presumably this paraphrase of Prawitz's restriction needs an amendment requiring that the two subproofs required as premisses for the rule of disjunction elimination [8, §20.2, R13] are also to be treated as "major premisses," but this is easily fixed. Indeed it seems to the writer in any event more natural to state the rule of disjunction elimination in the form "*r* is a direct consequence of $[p \vee q]$, $[p \supset r]$, and $[q \supset r]$," which makes its character as a sort of double-barrelled modus ponens more perspicuous, and makes it clear that the latter two premisses should be thought as "major premisses" for the rule, since from $[p \vee p]$, $[p \supset r]$, $[p \supset r]$, we can conclude *r* by disjunction elimination, and since *p* gives $[p \vee p]$, modus ponens is simply a special case.

Reference to the proof of Curry's Paradox stated earlier shows that the argument is attacked at a slightly different point: the proof is all right up until step 7, but in trying to prove step 8 we run afoul of the Prawitz Restriction, which forbids using step 5 as the major premiss for an elimination rule since step 5 has itself been obtained by an introduction rule. Nor does the Prawitz Restriction reject $[[p \& q] \supset r] \supset [q \supset r]]$, as did the Simple Restriction. But the Prawitz restriction admits proof of a theorem of the form $p \& {\sim} p$ and, therefore, the resulting system is consistent in neither the "weak" nor "strong" senses distinguished by Fitch and discussed below. For consider the concatenation of proof **I** followed by proof **II**. The only steps which are consequences by an introduction rule are steps **I**2, **I**5, **II**5, **II**8, **II**9, and **II**10. Step **I**2 ($=_{df}$ **I**3) is indeed a consequence by an introduction rule, but it does not serve as major premiss in going by an elimination rule from 1 and 3 to get 4, so the Prawitz restriction is satisfied, and inspection shows that none of the others among the steps **I**2, . . . , **II**10 listed just above is a major premiss for an elimination rule either. So, concatenation of **I** and **II** satisfies the Prawitz Restriction as stated by Fitch [11]. But now nothing prevents us from applying conjunction introduction to steps **I**5 and **II**10 to obtain a theorem of the form $p \& {\sim} p$. (Note incidentally that this fact does not lead to incoherence, since both **I**5 and **II**10 come by introduction rules and are hence ineligible for application of negation elimination).

But in spite of the fact that the Prawitz Restriction apparently avoids incoherence, semantical considerations make it still come too close for my comfort anyway, and I suspect, for Fitch's also, since it fails to satisfy even Fitch's criterion for "weak consistency," i.e., having no theorems of the form $p\&{\sim}p$. One could, I suppose, take the Wittgensteinian line that having such a contradiction doesn't hurt anybody, since one can't seem to do anything traumatic with it, and indeed Fitch [11, p. 264] says something similar in the last paragraph:

> The types of inconsistency that arise when S is weakly consistent but not strongly consistent are innocuous in the sense that, after all, the basic theorems that one wants for the foundations of mathematics remain provable and their denials remain unprovable.

I am inclined for reasons to be given later to think that it is still too early to give up the ship, although I don't think the Prawitz line helps us to keep it from sinking.

But this still leaves us with another, and for the present paper, final, possibility proposed by Fitch [11] under the title Restriction of Nonrecurrence. This seems to be the cleanest and smoothest of the restrictions available which guarantee coherence; Fitch [11, p. 258] states it as follows:

> A sentence which is an item of a proof or subproof cannot also be an hypothesis of a subproof which is an item of that proof or subproof.

Before commenting on the intuitive sense of this restriction, it is perhaps well to observe that it rules out such natural constructions as

1	p	hypothesis
2	p	hypothesis
3	p	1, reiteration
4	$p \supset p$	2–3, implication introduction
5	$p \supset [p \supset p]$	1–4, implication introduction,

which may seem odd. But nothing is lost, because implication introduction in Fitch [8] has two forms, of which the second is "$[p \supset q]$ is a direct consequence of q." So, we may get the theorem just above even in the presence of the Restriction of Nonrecurrence as follows:

1	p	hypothesis
2	$p \supset p$	1, implication introduction
3	$p \supset [p \supset p]$	1–2, implication introduction.

We still get the theorem required, but not by the first of the two subproofs just displayed.

One way of appreciating the intuitive sense of the Restriction of Nonrecurrence is to consider that any violation would have to have either Form A or Form B below:

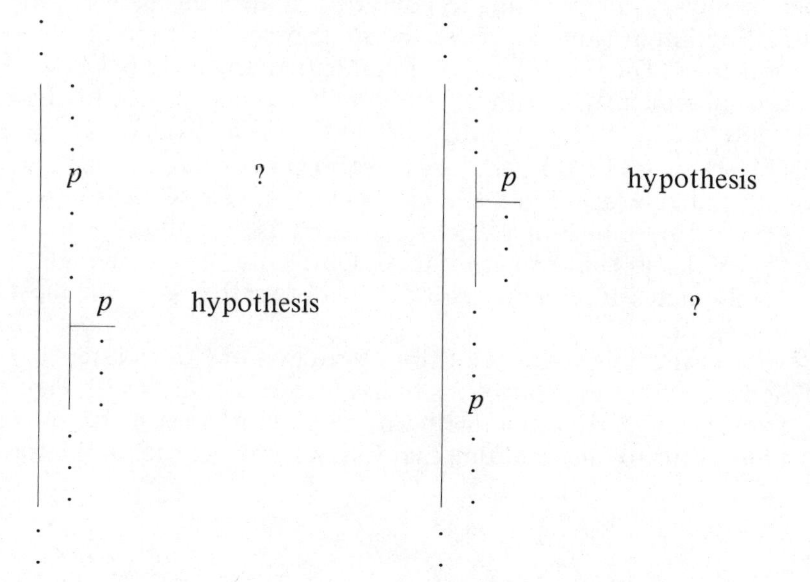

Form A Form B

In the case of Form A, we don't need the hypothetical subproof, since any work done in it could equally well be done by working from the first occurrence of p directly. In the case of Form B, (a) if the proof can be restructured so as to conform with Form A, then the remarks in the preceding sentence apply: we don't need the hypothetical subproof at all; and (b) if the proof in Form B can*not* be so restructured, then this is because the second occurrence of p depends (as a "resultant"?) on the hypothetical proof, as in the Curry argument near the outset of this paper. This means that the proof is bent on doing some sort of dirty work of Curry's kind, in which case we are glad to get rid of it. These considerations, together with some experiments in trying to find counterexamples, lead me to conjecture that the Special Restriction and the Restriction of Nonrecurrence should be demonstrably equivalent, or at least similar enough so that they could be shown with a little tinkering to amount to the same thing. So far, however, a quick and easy proof has eluded me.

Since we seem to be speaking intuitively at the moment, it may be somewhat illuminating to point out connections between the Restriction of Nonrecurrence and some other more familiar principles. The Curry Paradox as presented originally in Curry [5] and subsequently by others, was closely associated with the Law of Contraction, $[p \supset [p \supset q]] \supset [p \supset q]$, in the form $[[Y_q \epsilon Y_q] \supset [[Y_q \epsilon Y_q] \supset q]] \supset [[Y_q \epsilon Y_q] \supset q]$. We then define Y_q in such a way that the antecedent is satisfied, yielding the consequent $[Y_q \epsilon Y_q] \supset q$, which mysteriously gives us its own antecedent $[Y_q \epsilon Y_q]$ and hence an arbitrary q. Obviously there is enough uninhibited self-reference going on here to satisfy even the most narcissistic.

From the point of view of the Restriction of Nonrecurrence, however, the argument seems to have closer affinities with Pierce's Law $[[p \supset q] \supset p] \supset p$ than with Contraction. At least such connections occur to one in noting that Curry's argument takes the form

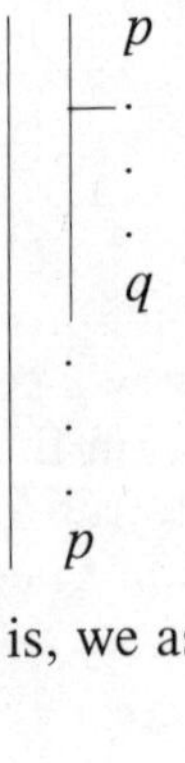

$$
\begin{array}{l|l}
 & \begin{array}{|l} p \\ \hline \cdot \\ \cdot \\ \cdot \\ q \end{array} \\
 & \cdot \\
 & \cdot \\
 & \cdot \\
 & p
\end{array}
$$

That is, we assume p, and then argue from the fact that p implies

something that p must therefore be true. Pierce's Law says that if the argument up to that point is sound, then p must be true. But Pierce's Law has the air of pulling one's self up by one's bootstraps and is of course not provable in Heyting's intuitionistic formal system, nor in Fitch's system (nor, we add in spite of an earlier promise, in the systems of connexive and relevance logic mentioned earlier).

II

We seem to be left at the moment with two viable alternatives: Fitch's original 1952 Special Restriction and the later Restriction of Nonrecurrence. I would now like to consider the role played in both cases by the introduction and elimination rules for negative implications.

In considering the matter, Fitch [11] offers a distinction between strong and weak consistency:

> A system will be said to be *weakly consistent* if it contains no theorem of the form '$p \& \sim p$'. On the other hand a system will be said to be *strongly consistent* if it contains no theorem 'p' such that '$\sim p$' is also a theorem of it.

Now if we add negative implication rules, then in the presence of a crucial assumption (which Fitch makes, and which we shall come to later on) **I**5 and **II**10 are theorems under either restriction, though their conjunction is not. Thus, both viable restrictions, in the presence of negative implication rules, lead to systems which are not "strongly inconsistent" (in the sense that they are weakly consistent: no theorems have the form $p \& \sim p$), but they are not strongly consistent either, since both **I**5 and **II**10 are theorems, though both restrictions prevent us from jamming **I** and **II** together into the same proof. The same is true of the formula $[[Y \epsilon Y] \supset \sim [Y \epsilon Y]] \supset \sim [[Y \epsilon Y] \supset \sim [Y \epsilon Y]]$ [11, p. 264]. Both this formula and its denial are provable in S with negative implication rules under either restriction, the system remaining weakly consistent but not strongly consistent, since the proofs of the formula in question and its denial cannot be laid end-to-end without violating one restriction or the other, as the reader may check. And no doubt many other examples of such goofy formulas could be cooked up with a little ingenuity.

So both restrictions allow the existence in S of a batch of theorems *p,* such that $\sim p$ is also a theorem. Like a cloud of gnats, these buzz around in front of our eyes, though they don't seem to do too much harm, in that they are Specially or Nonrecurrently restricted from mating, so to speak, and begetting chaos.

Now we might say that the cloud of gnats is harmless, a mere nuisance that doesn't really keep us from getting on with our work except that it is irritating to have to keep brushing them away. But from a semantical point of view, they look more menacing, since they seem to preclude giving S any coherent interpretation. Observe that in addition to **I** above we have

III	1	$[Y_{\mathbf{r}} \epsilon Y_{\mathbf{r}}]$	hypothesis
	2	$Y_{\mathbf{r}} \epsilon \hat{x}[[x \epsilon x \supset \mathbf{r}]$	1, repetition, abbreviation
	3	$[Y_{\mathbf{r}} \epsilon Y_{\mathbf{r}}] \supset \mathbf{r}$	2, abstraction elimination
	4	$\mathbf{r}$	1, 3, implication elimination
	5	$[Y_{\mathbf{r}} \epsilon Y_{\mathbf{r}}] \supset \mathbf{r}$	1–4, implication introduction.

Both restrictions rule out deriving **r** (i.e., $\sim[p \supset p]$), from **I**5 and **III**5, since both prevent laying proofs **I** and **III** end-to-end. But neither **I** nor **III** violate either restriction, so (1) *S is in neither case closed under modus ponens for the horseshoe.* What then does the horseshoe mean?

Secondly, both **I**5 and **I**10 are theorems, but with either restriction their conjunction is not a theorem, so (2) *S with either restriction is not closed under conjunction introduction.* What then is the ampersand to mean?

Finally, we have theorems *p* such that $\sim p$ is also a theorem, so (3) *the truth of p* (if all theorems are to be true) *does not preclude the truth of* $\sim p$. Ordinarily, at least, the point of making such dichotomies as true-false, valid-invalid, provable-unprovable, and the like, is that they are pairwise mutually exclusive, even if, with Fitch and the intuitionists, we wish to allow for some No Man's Land between truth and falsity.

These difficulties can, however, all be cleared up if we depart slightly from or perhaps just reorganize a little, the classical way of looking at formal systems. Though not put in exactly these words the tenor of §4 of Fitch [8] suggests that we must first define *proof* in some way, then define *theorem* as a formula which is a propositional item of a categorical proof, and finally define the *system* S as the set of those formulas for which there are proofs, i.e., the set of theorems.

However, there is nothing but stale custom which *requires* us to think of matters so. We might instead say that axioms, and the positive and negative introduction and elimination rules allowed (i.e., all but negation introduction), define a set of distinct sequences of formulas, each of which satisfies (say) the Restriction of Nonrecurrence and that in a sense each sequence constitutes a system of logic in its own right.

Let us at any rate contemplate for the sake of argument the denumerable set **F** of finite proof sequences S_i, each of which satisfies the conditions (1) every item is either an axiom, according to R1 or R2 [8, §20.2], or else a direct consequence of predecessors by one of the rules R3–R23, with R13, for disjunction elimination, amended as suggested above, or else a direct consequence by the introduction and elimination rules for negative implications; and (2) S_i in toto satisfies the Restriction of Nonrecurrence.

We note first that each S_i is strongly consistent in Fitch's sense as quoted above, i.e., consistent with respect to negation in the usual sense, which is essentially the content of Fitch's consistency proof in [8, §20] as amended for inclusion of negative implication rules in [11]. Moreover, each of the S_i fails to have the disagreeable semantical features mentioned above: each S_i is closed under implication elimination (modus ponens), so that the horseshoe may be given the "if . . . then —" sense motivated by its introduction and elimination rules; similarly for conjunction introduction. Each S_i is also closed under all the other rules of the system S_i, including rules for negative implication and the three rules for negation.

How do the systems S_i in **F** differ? The answer sheds light I think on Fitch's claim that the contradictions in a system which is weakly but not strongly consistent are "innocuous." In effect, such crackpot constructions as proofs **I** and **II** present a trilemma to S_i, which can be roughly phrased: "As you lengthen yourself, of proofs **I** and **II** you may pick one or the other or neither; but the rules forbid you to pick both."

If we may draw a slightly different picture, we may imagine a clearheaded, sober-minded mathematician setting out to construct for us an S_i which has as its conclusion some truth *T*, provable with the apparatus thus far developed. He starts steadily enough, applying rules to identities and nonidentities provided by R1 and R2, then other rules to the results, and so on. Then at some point a wild-eyed logician, for God knows what perverse reason, insists that he write down a proof of **I** or a proof of **II**. The mathematician, though nettled by the insistence, gives in grudgingly, whereupon the logician asks "which one?" Whereupon the mathematician retorts "It couldn't make less difference; you choose it, since it won't affect matters anyway. It won't help with my proof, but it won't hurt either, since the Restriction of Nonrecurrence guarantees that if I pick one, I won't get the other. Whichever is chosen will just sit there twiddling its thumbs. For myself I'd prefer to add neither **I** nor **II** to the proof I'm working on—I'm just letting you pick one because I'm trying to be easy to get along with, anyway."

And now a plausible move suggests itself for obtaining something like a single system of logic which, though nonconstructive, is clearly consistent. That is, we take as theorems of the system S the set of formulas common to all sequences S_i. This move will exclude some formulas that occur in **I** but not in **II**, i.e., we don't get both 5′ and 10′. In fact, we get neither.

And there we have it all: consistency comes from the fact that neither of the theorems our mathematician in the paragraph above objected to are forthcoming, and power comes from the fact that it is all so prodigiously nonconstructive. But for anyone who has accepted "Carnap's rule" in the first place, we are—if we are really honest with ourselves—so far away from the Hilbert program, that leaping ϵ miles beyond the last ϵ-number shouldn't bother our consciences too much.

References

(*JSL* abbreviates *The Journal of Symbolic Logic.*)

[1] Anderson, Alan Ross, and Belnap, Nuel D., Jr. 1975. *Entailment: the logic of relevance and necessity,* Princeton: Princeton University Press.

[2] Angell, R. B. 1962. A propositional logic with subjunctive conditionals. *JSL* 27: 327–43.

[3] Belnap, Nuel D., Jr. 1973. Restricted quantification and conditional

assertion. In *Truth, syntax, and modality,* ed. Hugues Leblanc, pp. 48–75. Amsterdam: North Holland.
[4] Church, Alonzo. 1956. *Introduction to mathematical logic.* Princeton: Princeton University Press.
[5] Curry, Haskell B. 1942. The inconsistency of certain formal logics. *JSL* 7: 115–17.
[6] Fitch, Frederic Brenton. 1938. The consistency of the ramified *Principia. JSL* 3: 140–49.
[7] ——. 1942. A basic logic. *JSL* 7: 105–14.
[8] ——. 1952. *Symbolic logic, an introduction.* New York: Ronald Press.
[9] ——. 1953. Self-referential relations. *Proceedings of the eleventh international congress of philosophy* 14: 121–27.
[10] ——. 1963. The system CΔ of combinatory logic. *JSL* 28: 87–89.
[11] ——. 1969. A method for avoiding the Curry paradox. *Essays in honor of Carl G. Hempel,* pp. 255–65. Dordrecht: Reidel.
[12] ——. 1971. Propositions as the only realities. *American philosophical quarterly* 8: 99–103.
[13] Kemeny, John G. 1948. Models of logical systems. *JSL* 13: 16–30.
[14] McCall, Storrs. 1966. Connexive implication. *JSL* 31: 415–33.
[15] Meyer, Robert K., and Dunn, J. Michael. 1969. E, R, and γ. *JSL* 34: 460–74.
[16] Prawitz, Dag. 1965. *Natural deduction.* Stockholm: Almqvist & Wiksell.
[17] Russell, Bertrand. 1903. *The principles of mathematics.* London: Allen & Unwin.

10: NUEL D. BELNAP, JR.

Grammatical Propaedeutic*

The principal aim of this essay is to convince the reader that it is philosophically respectable to "confuse" implication or entailment with the conditional and, indeed, philosophically suspect to harp on the dangers of such a "confusion." The suspicion is that such harpists are plucking a metaphysical tune on merely grammatical strings.

Our strategy is as follows. First we lay out as clearly as we can the grammatical background—call it "logical grammar"—against which the distinction of conditional from implication should be made, for here as elsewhere confusion is only admissible when clarity is possible. This effort terminates in a table laying out a variety of grammatical possibilities for certain key philosophical concepts: truth, negation or falsity, conjunction, conditional, and entailment. Second, we discuss the logical grammar of truth, negation or falsity, and conjunction in order to elicit easy-to-overlook features of the application of logical grammar to English, summarizing our views in eight theses. Lastly, having prepared the way (and after a brief word about quantifiers), we discuss the grammar of conditional and entailment.

A1. Logical Grammar

By a logical grammar we mean one of the sort that guides the thinking of most logicians. Such a grammar is a more or less language-independent family of grammatical categories and rules clearly deriving from preoccupation with formal systems but with

*This essay is offered in affection and respect to Professor F. B. Fitch, who first introduced me to the fascination of logic. A version will also appear as an appendix to [1]. The research was partially supported by NSF Grant No. GS 28478. Thanks for suggestions are due to D. Grover, T. Parsons, and R. Thomason; and, as always (but this time posthumously), to A. R. Anderson, with whom I shared the pleasure and profit of logical lunches with Professor Fitch at My Brother's Place.

at least prospective applications to natural languages. A logical grammar is accordingly one which is particularly simple, rigorous, and tidy, one which suppresses irregular or nonuniform or hard-to-handle details (hence differentiating itself from a "linguistic" grammar whether of the MIT type or another), one which idealizes its subject matter, one which by ignoring as much as possible leads us, perhaps wrongly, to the feeling "Aha! That's what's *really* going on!" Among such logical grammars we think a certain one lies, unexpressed, in the back of most logicians' heads. Our task is to make it explicit, not in order to criticize it–indeed though it would be dangerous to suppose it eternally ordained, we think very well of it–but because only when it is brought forth in all clarity can we sensibly discuss how it ought to be applied. The version here presented derives from Curry and Feys [3] and Curry [2].

Of course, here and hereafter application of any logical grammar will be far more straightforward and indisputable in the case of formal languages than in the case of English; for in the case of the former, possession of a tidy grammar is one of the design criteria. But English is like Topsy, and we should expect a fit only with a "properly understood" or "preprocessed" English–the preprocessing being to remove, of course, the hard cases. So, read on with charity.

Logical grammar, as we understand it, begins with three fundamental grammatical categories; the sentence, the term, and the functor. The first two are taken as primitive notions, hence undefined; but we can say enough to make it clear enough how we plan to apply these categories to English and to the usual formal languages.

By a *sentence* is meant a declarative sentence in pretty much the sense of traditional grammar. Some part of logical opinion would probably exclude traditional sentences like "It is a dog," which are fit only to receive a truth value in context instead of an outright truth value; but in order to cast our net as widely as possible, we generously allow such candidates full status as sentences in our logical grammar. On the formal side, we intend for the same reason to include "Fa" even when the denotation of the individual constant "a" is possible-world- or context-dependent (in a modal or tense logic), and even "F*x*," where "*x*" is squarely a free variable.

Traditional grammar gives us less help in articulating the concept of a *term,* although the paradigm cases of both traditional nouns and logical terms are proper names such as "Wilhelm Ackermann."

Some more examples of terms from English and near-English:

Terms	*What some linguists call them*
Wilhelm Ackermann	proper noun
the present king of France	noun phrase
your father's moustache	noun phrase
triangularity	(abstract) noun
Tom's tallness	(abstract) noun phrase
that snow is white	that clause: factive nominal; nominalized sentence
what John said	factive nominal; nominalized sentence
his going	gerund; nominalized sentence
he	pronoun

We add some further examples from formal languages.

$3+4$	closed term
x	variable
$3+x$	open term
$\{x: x \text{ is odd}\}$	set abstract
ιxFx	definite description

What the logical grammarians contrast with these are so-called *common nouns* such as "horse," as well as plural noun phrases such as "Mary and Tom," mass nouns such as "water," and indefinite noun phrases such as "every man" and "a present king of France." And although we take the category of terms to be grammatical, it is helpful to heighten the contrast by a semantic remark: the terms on our list purport—at least in context or when fully interpreted ("assigned a value")—to denote some single entity, while "horse" and "Mary and Tom" do not (*pace* those who want to calculate with individuals). Perhaps the common noun is the most important English grammatical category not represented anywhere in logical grammar. Contemporary logicians, including us, uniformly torture sentences such as "A horse is a mammal," which contain common nouns, into either "The-set-of-horses is-included-in the-set-of mammals," or "For-anything you-name, if it is-a-horse, then it is-a-mammal," where the role of the common noun "horse" is played by either the term (our sense) "the-set-of-horses" or the predicate (see below) "— is-a-horse."

So much for sentence and term. The third fundamental grammatical category is constituted by the *functors.* By a *functor* is meant a way of transforming a given ordered list of grammatical entities, i.e., a list the members of which are terms, sentences, or functors, into a grammatical entity, i.e., into either a term, a sentence, or a functor. That is to say, a functor is a function—a grammatical function—taking as inputs (arguments) lists of items from one or more grammatical categories and yielding uniquely as output (value) an item of some grammatical category. For each functor, as for any function, there is defined its *domain,* that is, the set of its input lists, and its *range,* which is the set of its outputs.

We are really only interested in what we will eventually call pure elementary functors, but in order to appreciate the limitations we lay on ourselves, and so as not to fossilize our grammatical (philosophical?) imaginations, it will be useful to survey the functorial landscape by means of some definitions and examples. Those uninterested in landscapes should skip four paragraphs to the displayed table of pure elementary functors.

Given the domain of a functor, it can happen that the length of every input list is the same, or it can happen that there are input lists of different lengths. In the first case we say that the functor is *fixed.* For example, the "if-then" functor is fixed because it always takes as input a list of length two. Fixity will be part of the definition of *elementary,* but there are important cases of nonfixed (variable) functors. For example, consider the functor in set theory which from a list of entity names produces a name of the set of those entities, i.e., the functor represented by the notation "$\{x_1, \ldots, x_n\}$," where n is allowed to vary. This functor is not fixed. Of course one could define "$\{x_1, \ldots, x_n\}$" as representing a different functor for each n; each such functor would then be fixed. But this is a trivial point, and does not remove the possibility of someone else presenting the grammar of the language of set theory using the nonfixed functor we described.

So much for fixity. As we have defined them, functors can take other functors as arguments. The importance of such functors is enormous; e.g., English "very" should probably be construed as taking a predicate (see below), e.g., "— is red," into a predicate, e.g., "— is very red." But we shall not be talking about such higher level functors, a fact we emphasize by calling a functor *first level* if its input lists are wholly confined to sentences and terms. The functors represented by "if — then —" and "— + —" are first level in this sense, as is also "— believes that —." In order to avoid considering examples like the last, where an input list can be a mixture of terms and sentences, as well as examples like quotation marks, which in English can take either a term, as in " 'Tom'," or a sentence, as in " 'Tom is tall'," we call *pure* those among the first-level functors which have input lists consisting always either wholly of terms or wholly of sentences, and when furthermore, the output is either always a term or always a sentence. When we use examples from English, they are charitably to be construed as pure first-level functors.

We turn now to the last subdivision of functors. By a *substitution* functor we mean a functor for which there is a nonempty pattern (or possibly a family of patterns) of words with blanks, the blanks being ordered, such that for each input list in the domain of the functor, the output can be obtained by substituting the elements of the input list, in order, into the blanks. "If — then —" and " — + —" represent substitution functors. Ordinary negation in English is *not* a substitution functor: the passage from "Snow is white" to its ordinary negation, "Snow is not white," cannot be accomplished by substituting the former into a blank. In contrast, the logician's negation *is* a substitution functor: one can pass from "Snow is white" to "It is not the case that snow is white" by substitution into "It is not the case that —." Substitutionality will be part of our definition of *elementary*.

A substitution functor is defined once one knows (1) the pattern of words with (ordered) blanks defining the pattern of substitution, and (2) the domain of inputs. It is therefore often convenient to speak as if the pattern of words with blanks ordered from left to right were itself the functor, especially in giving examples. But in such cases one must keep in mind that the domain of the functor then has to be either separately specified or gathered from context. For example, we will allow ourselves to say "the functor '— + —'," with the understanding that only number terms are to be allowed as inputs.

We can now define the elementary functors: an *elementary functor* is a fixed first-level substitution functor. Elementary functors do not have to be pure, but we shall confine our attention to those that are; and evidently there are four kinds of pure elementary functors, since the inputs can be either terms or sentences, while the outputs may vary in the same way. The four kinds of pure elementary functors are exhibited, named, and exemplified in the table below:

Inputs	*Output*	*Name*	*Examples*
Terms	Term	Operator	— + —; —'s father
Terms	Sentence	Predicate	— < —; — is nice
Sentences	Sentence	Connective	— and —; John surmises that —
Sentences	Term	Subnector	'—'; that —

(We note in passing that individual quantification is elementary, but not pure: "(∃ —) —" is a fixed first-level (but not pure) substitution functor, taking variables (a species of term) in the first blank and sentences in the second, and producing a sentence.)

Subnectors—the last entry in the table—are seldom mentioned and seldom occur in formal languages, but they are of particular interest for the present discussion. A subnector is a way of converting a sentence into a term; the examples given in the table above are the two subnectors of most importance for us in English. We shall call them, respectively, the *that subnector,* and the *quote subnector.* (We shall use single quotes here, reserving double quotation to be used informally in the language we are using. Note that we always suppress double quotes (used quotes) from displays, and that single quotes, in displays or not, are always mentioned and never used.) Observe that putting a sentence into the blank in either

'—'

or

that —

invariably results in a term, e.g.,

'Tom is tall',

or

that Tom is tall.

These thus satisfy the definition of *subnector.* The first display above may be taken as the name of the sentence, the second, perhaps, as the name of a proposition.

Terms in logical grammar may usefully be subdivided into finer grammatical categories. For example, number terms, physical object terms, set terms, city terms, etc., would all very likely prove useful categories. In giving a category of terms a name, we find it most convenient to use something suggesting the domain of entities which will be taken as the denotata of these terms, but the idea is that in logical grammar the category shall be defined independently of semantics. (This may well be impossible for English.)

Of special interest in what follows will be the categories of "sentence term" and "proposition term." One sort of *sentence term* is obtained as the result of applying the quote subnector " '—' " to a sentence, and one sort of proposition term as the result of applying the that subnector "that —" to a sentence. Examples:

Sentence term: 'Tom is tall'
Proposition term: that Tom is tall

But we would naturally suppose that "the last sentence in *Paradise Lost*" would also be counted as a sentence term, and that "what he said" or "Euclid's first proposition" or perhaps even "Tom's tallness" would count as proposition terms.

Whenever there is a subdivision of terms, it is possible and usually useful to subdivide the various functors involving terms according to what sort of terms they take as input and what sort they produce as output. For example, " — + —" is a number-into-number operator: its inputs must be number terms, and its output is always a number term. And "— < —" is a number predicate: its inputs must be number terms. Also, the subnector "the probability that —" is a number subnector, since its output is a number term.

Of special interest in logical theory are the functors whose inputs or outputs are restricted to either sentence terms or proposition terms. A *sentence predicate* is one in whose blanks you can put only sentence terms, and a *proposition predicate* one in which only proposition terms can be put. *Sentence operator, proposition operator, sentence subnector,* and *proposition subnector* are similarly defined. Examples:

— is true

is a sentence predicate or a proposition predicate, according as the separately specified domain is restricted to sentence terms or proposition terms. As a sentence predicate, "— is true" yields " 'Tom is tall' is true," and as a proposition predicate it yields "That Tom is tall is true." It has often been pointed out that "— is true" cannot be a connective, since "Tom is tall is true" has too many main verbs.

'— is true'

is a sentence operator when the domain is restricted to sentence terms, yielding, e.g., " ' 'Tom is tall' is true'." If its domain is restricted to proposition terms, it would be a proposition-to-sentence operator, an eventuality we ignore.

that — is true

is a proposition operator when the domain is restricted to proposition terms, giving, e.g. the term "that that Tom is tall is true."

Again there is a sentence-to-proposition operator in the vicinity which we shall ignore; but more seriously, note that "that — is true" is also a connective, with a sentential domain, since one can put a sentence in the blank to get, e.g., the sentence "That Tom is tall is true." This possibility arises in the case of "that — is true," but not " '— is true'," because the scope of "that" in English is sometimes ambiguous where the scope of a pair of quotes would not be. In order to reduce side remarks indicating domains, we shall sometimes use parentheses to indicate the scope we have in mind for "that;" thus writing "that (— is true)" for the operator and "(that —) is true" for the connective.

Lastly, the quote subnector and the that subnector are prime examples respectively of sentence subnectors and proposition subnectors. Two further examples:

Sentence subnector: 'if — then —'
Proposition subnector: that if — then —.

If one puts sentences in the blanks, one produces a term of the appropriate sort, thus satisfying the defining conditions.

Before applying this analysis, let us summarize our logical grammar. (1) A grammatical entity is either a sentence, a term, or a functor. (2) A functor is a mapping from lists of grammatical entities into grammatical entities. (3) There are many kinds of functors, among which we isolate the pure elementary functors. (4) There are four kinds of pure elementary functors: predicates, operators, connectives, and subnectors. (5) The isolation of the categories sentence term and proposition term induces a classification of predicates, operators, and subnectors into sentence predicates vs. proposition predicates, etc.

A2. The Table

Now for the application. Whenever we approach a net of concepts with the idea of trying to get clear on them, a good question to ask is, How should we parse the fundamental terminology? Our goal is the net involving the concepts of entailment or implication, and the conditional. But let us make haste slowly by simultaneously raising the question for other key logical concepts such as truth, negation, and conjunction.

What the foregoing gives us is an account of the range of possibilities for parsing entailment (we'll stick to "entailment" even though our remarks apply exactly to implication) and the conditional, truth, negation-falsity, and conjunction. We should expect to find in the vicinity of each of these concepts each of the following: a predicate, a connective, an operator, and a subnector; and usually both a proposition variety and a sentence variety for each. A selection of possibilities is laid out in the following table.

Truth		
Operator:	that (— is true),	'— is true'
Predicate:	— is true	
Connective:	(that —) is true,	'—' is true
Subnector:	that ((that —) is true),	' '—' is true'
Negation or falsity		
Operator:	that (— is false)	
Predicate	— is false	
Connective:	(that —) is false	
Subnector:	that ((that —) is false)	
Conjunction		
Operator	that (— is true and — is true)	
	—⌢'and'⌢—	

Predicate:	— is true and — is true	
Connective:	— and —	
Subnector:	that (— and —),	'— and —'
Conditional-entailment	(with "if" and then with "entails")	
Operator:	that (if — is true then — is true)	
Predicate:	if — is true then — is true	
Connective:	if — then —	
Subnector:	that if — then —,	'if — then'
Operator:	that (— entails —)	'— entails —'
Predicate:	— entails —	
Connective:	(that —) entails that — '—' entails '—'	
Subnector:	that ((that —) entails that —)	

In every place in which there is a blank for terms, the domain could be restricted to either proposition terms or to sentence terms. And in every place in which a "that —" is displayed, one could have a variation with quotes instead; but we have presented the quote versions in only a selection of the cases above.

A3. Eight Theses

Using this table as a basis, we arrive at eight theses.

A3.1. Logical Grammar and Logical Concepts

Our first and basic thesis is perhaps the most important of all. It is that logical grammar is in fact applicable to English in the sense that we can identify various constructions in English as satisfying the conditions defining a certain kind of functor. For example, we note that "if — then —" is a connective, and that "— entails —" is a predicate, the former requiring sentences and the latter requiring terms, an observation of importance to anyone who wishes to be grammatical. Indeed, we think that logical grammar should be a part of the curriculum of every introductory logic course. So:

THESIS 1: Logical grammar is applicable to English.

We next remark that the table both presupposes and leads to the claim that it makes sense to talk about a concept prior to specifying its grammatical form. For example, by "negation-falsity" we mean to refer to that generalized concept of naysaying (we stretch for a generic term) which has not yet decided what grammatical clothes to wear. Of course, logicians customarily use "negation" for a connective and "falsity" for a predicate—although using the latter ambiguously for both a propositional and sentential predicate—but this should not blind us to the existence of a topic of which these two are specializations. Similarly, although a certain school of logicians systematically uses "conditional" for the connective and "implication" for the predicate, it is no confusion to suggest that there is a single topic here. Analogously, there is a concept of mass in physics before one has decided upon its grammatical role in a tidied up reconstruction: whether to construe it as carried in the language of physics by an equivalence predicate, a physical object-into-number operator, four-place predicates representing interval or ratio similarities, or what. The same remark holds for time and other topics to which we are accustomed to refer by means of an abstract noun; for unless there is something antecedently there, it makes no sense to ask the extremely important question as to the most fruitful grammatical analysis of the concept (or family of concepts) in question. We are accordingly led to the following

THESIS 2: A concept (or family) can be isolated without specifying its grammar; and, accordingly, one can question the advantages and disadvantages of carrying it in various grammatical forms.

A3.2. A Question of Fit

One can also ask about the fit between English and logical grammar with respect to a given set of concepts.

A3.2.1 Simplest Functors

We consider the various logical concepts in turn, pointing out in each case that functor in the vicinity of a given concept which is most easily expressed in English.

Truth: If we disregard the null functor, the grammatically simplest pure elementary functor in the vicinity of truth is a predicate. We also note that English is indifferent as to whether the truth predicate is construed as a sentential predicate or as a propositional predicate; both, in English, are equally grammatical.

Negation-falsity: Exactly the same comment applies here.
Conjunction: The simplest functor in the vicinity is a connective.
Conditional and entailment: The simplest functor involving "if" is a connective, while the simplest involving "entails" is a predicate.

We may summarize by means of the banal
THESIS 3: English permits a given logical concept to be expressed by functors which differ in simplicity.

A3.2.2. More Complex Functors

We continue our investigation of English grammar from the point of view of logical grammar by looking at a selection of more complex functors attaching to the various logical concepts. First as to truth and negation-falsity: Every substitution connective for either truth or negation-falsity seems to involve prenominalization via either the quote subnector of the that subnector. So if one wants a systematic way of negating or of truifying in English, one seems required to use one of these functors, as in "that A is false" or "that A is true." Or perhaps in the case of truth, we might use the connective "Truly, —" as in "Truly, snow is white," or even the "null" connective "—" as a truth connective without prenominalization via a subnector; but we cannot happen to think of anything analogous in the case of falsity. Of course English does have at least a partial (i.e., perhaps not everywhere defined) connective for negation-falsity that does not involve prenominalization: we mean the connective that takes "Snow is puce" into "Snow is not puce," "Snow is not puce" into "Snow is puce," etc. But this connective is not elementary; in particular, it is not a substitution functor. It gives us no uniform way of negating A, especially when A is a complex sentence.

Now for conjunction. If we wish to have a conjunction predicate, we seem to need to insert "— is true." This is perhaps not surprising; but it is remarkable that we need to involve "— is true" in finding an elementary propositional conjunction operator. But let us see first how to find a substitutional sentential conjunction operator. We are given as inputs the sentence terms " 'Snow is white' " and " 'Tom is tall'," and wish to produce as output a sentence term to be taken as naming what amounts to a conjunction of what the two input terms name; i.e., a single sentence which implies exactly those sentences jointly implied by what the input terms name. So, we want " 'Snow is white and Tom is tall'." Obviously we have just done what is required; but notice that we have not employed a substitution functor in so doing. Using "S" and "T" in the obvious way, the passage from " 'S' " and " 'T' " to " 'S and T' " is not and cannot be accomplished by substitution. There are ways out, perhaps the simplest being to use concatenation: "—⌢'and'⌢—" applied to " 'S' " and " 'T' " yields " 'S'⌢'and'⌢'T' ," which is indeed a name of the conjunction "S and T" of "S" and "T". And note—this is relevant to our next point—that this works even if one puts in the blanks a sentence term in the wider sense, e.g. "the first sentence in *Paradise Lost*." (But is it English?)

Now for the *propositional* conjunction operator. We are given as inputs the proposition terms "that S" and "that T," and wish to produce a proposition term to be taken as naming what amounts to a conjunction of what the two input terms name; so we want "that S and T." But, again, we have not employed a substitution functor. And in this case the only way out that occurs to us employs presententialization via truth: "that — is true and — is true," yielding "that that S is true and that T is true" by substitution, and naming what amounts to a (propositional) conjunction of what "that S" and "that T" name. Furthermore, consider the problem of forming a name of the proposition amounting to the conjunction of what Peter said and what Paul said. We seem in the presence of sentence terms in the wider sense to have to settle for "that what Peter said is true and what Paul said is true," thus involving "— is true" to keep the grammar straight by presententializing the proposition terms. Note that "what Peter said and what Paul said" is grammatical, but that it doesn't name the conjunction required, being instead a plural term and thus falling outside the scope of logical grammar. Of course in technical English "the conjunction of — and —" would do; but few would understand "I deny the conjunction of what Peter said and what Paul said" without an explanation. So if we want to remain with ordinary English, "true" seems the only way.

Similar remarks apply to conditional and entailment: If we wish to have a predicate involving "if," we are going to have to presententialize by sprinkling in some occurrences of "true," and if we want a connective in the vicinity of entailment, we shall have to prenominalize by inserting either some occurrences of "that" or some quotes. We may summarize these considerations in the following two part

THESIS 4: (a) Given a logical concept, we can always find a way to carry it in English as any functor we like, provided we are willing to presententialize and prenominalize by means of a sufficient number of instances of "is true" and "that." (b) Often the only or at least the easiest way to carry a concept as a given functor is to presententialize or prenominalize by "is true" or "that."

A3.3. Parsing Logical Concepts

The foregoing constitutes some more or less factual observations concerning the relations between English and logical grammar. We now want to turn more directly to the question, How should we parse the terminology governing the key concepts of logic?

First let us ask Pilate's question, What is Truth? In English "truth" is sometimes a common noun, as in "We take these truths to be self-evident," and sometimes a term, as in "Truth is beauty." But we systematically abjure common nouns in logical grammar, and although we do sometimes want a term "Truth" or "the True" as a name of one of the truth values, and although Frege seems to have deployed "Truth" or "the True" as a sentence, the usual problem is as to whether we want the concept of truth carried by a predicate or a connective–and if a predicate, whether sentential or propositional. Tarski's answer is well known: the important locution in the vicinity of "truth" is the truth predicate,

— is true.

Furthermore, Tarski takes this predicate to be a sentence predicate, taking in its blank only sentence terms. Others have argued that the truth predicate is a proposition predicate, taking proposition terms in its blank. English as she comes certainly accepts both:

'Snow is white' is true

and

(that snow is white) is true

are equally grammatical. The philosophical question of which (if either) truth predicate to take as the foundation for a theory of truth must therefore be answered on other grounds. Which is our next (obvious) point:

THESIS 5: Sometimes English is indifferent as to grammatical form, in which case the choice of functor with which to carry a concept must be made on extragrammatical grounds.

Our further and more important point: the fact that "— is true" is the simplest functor in the vicinity of truth (Thesis 3) does not entail that one should so parse truth in laying the foundations of an adequate theory; the structure of English may turn out to be a bad source for intuitions guiding the construction of a fruitful theory of truth. It is open to the logical theorist, given "that Tom is tall is true," to take as immediate constituents (1) "Tom is tall" and the connective "that — is true," rather than (2) "that Tom is tall" and the predicate "— is true," and thus opt for some other functor more difficult to express in English (Thesis 4). Of course any sensible English grammar will proceed according to (2); but that should not dictate how we apply logical grammar.

For consider negation-falsity in parallel with truth. Both have associated with them an operator, a predicate, a connective, and a subnector, and in each case the predicate is the simplest substitutional functor. But when it comes to founding a theory on a choice of fundamental grammar, the astonishing fact is that logicians almost uniformly choose differently in these two cases: they carry truth by a sentence predicate, and negation-falsity by a connective. The most interesting and fruitful theory of truth is taken to be Tarski's semantic one, not the theory of the null connective or of "Truly, —," while the most interesting and fruitful theory of naysaying would by most be taken to be the theory of the negation connective, read in English as "that — is false" or "it is not the case that —" or "that — is not the case." We are not challenging the decision. Our only point is that in English grammar truth and falsity are on all fours, so that other considerations must be invoked in support of the logician's decision to treat them differently.

Of crucial importance, both here and later, is the fact that the presence of "that" in the falsehood connective, "that — is false," or its cousins mentioned above, does not make the typical logician feel that he must construe outputs from this connective, such as "that Tom is tall is false," as somehow "about" a proposition, that Tom is tall, or as somehow implicitly metalinguistic.

In fact, given

It is not the case that Tom is tall,

logicians uniformly take the "immediate constituents" to be

Connective: "It is not the case that —"

and

Sentence: "Tom is tall."

They do not take the immediate constituents as

Predicate: "It is not the case —"

and

Term: "that Tom is tall,"

because, presumably, they do not wish to "commit" themselves to the existence of some entity named by the term "that Tom is tall." Again it is to be pointed out that it is theoretical considerations and not English grammar which dictate the choice, for certainly any sensible English grammar will take "that Tom is tall" to be an immediate constituent of "It is not the case that Tom is tall" or of its underlying form "That Tom is tall is not the case." But that should not dictate how we apply logical grammar. We are thus led to the following

THESIS 6: In contrast to the situation envisaged in Thesis 4, sometimes English has fixed ideas about grammatical form: some English constructions definitely and uniformly require terms, and some sentences. But in formulating the theory of a given logical concept, although we should of course be grammatical, it is not required that we slavishly take the details of English grammar as a sure guide to the most fruitful way to proceed.

A3.4. Reading Formal Constructions into English: the Role of "True" and "That"

We now wish to apply these considerations to the problem of finding readings from a formal language into English, starting with formal negation—the curl—as our example. We know from Thesis 3 that every substitution connective for either truth or negation-falsity (with the mentioned odd exceptions for truth) seems to involve prenominalization via either the quote subnector or the that subnector. So if one wants a systematic way of reading "∼—" in English, one seems required to use one of these functors, as in "It is not the case that —." And the same holds for a truth connective, "T(—)"; we have to say something like "that — is the case."

What follows? We think nothing does. We think this is a sheer grammatical fact about English and has no philosophical significance. We think there is nothing "conceptual" about the fact that "Not Tom is tall" or "It is not the case Tom is tall" or "Tom is tall not" are bad English grammar, while "Tom is tall and Mary is bald" is as good a sentence as "That Tom is tall and that Mary is bald" is a bad sentence. English has a convenient substitution connective for conjunction which does not take a detour through subnectors, but it does not appear to have a convenient substitution connective for either truth or falsity except those involving subnectors. It is accordingly easier to find a reading into English for formal conjunction than it is to find one for formal negation. There's a fact for you, and a mere fact.

We think the matter sufficiently important to illustrate with respect to another example, this time involving conjunction as a propositional operator. To give point to the illustration, we first remark that on nongrammatical grounds we think the theory of the conjunction operator an interesting one. It increases our understanding of conjunction to have it explained that given two propositions a and b, their conjunction is (the proposition which is) the greatest lower bound of a and b with respect to the ordering of propositions by the relation of propositional implication. In other words, the conjunction of propositions a and b is the weakest proposition implying both a and b. So we introduce notation "$a \wedge b$", say, for the conjunction of a and b. But how now are we to give a simple and uniform reading of "$a \wedge b$" in English? If we mind our grammar, the only way seems to be something like "that both a is true and b is true;" or "that a obtains and b obtains." But this reading appears to involve us with the concept of truth, although it nowhere plays a role in the theory of conjunction sketched above. And if we have been brought up by Tarski, it appears to be metalinguistic, although our theory was not.

We may, however, conclude from the foregoing that these appearances are illusory; we are in this case using "— is true" or

"— obtains" only because English "happens" not to have a simple propositional conjunction operator not involving (something like) these words. It might have been that English was possessed of a conjunction operator, say "— et —," so that what Peter said et what Paul said would be the weakest proposition implying both what Peter said and what Paul said. And–we think–this is a might-have-been which would not have altered the deep conceptual structure of English, whatever that is. We are therefore at liberty to treat our forced reliance on "is true" as a philosophically uninteresting feature of English.

Someone who abhors propositions is of course also at equal liberty to abhor our entire theory of propositional conjunction; we do not mean to be arguing for it just here. Our only point is that such a person is not entitled to buttress his arguments with grammatical fact unless he is also prepared to argue noncircularly that these facts are of philosophical interest. Thus:

THESIS 7: There are important grammatical constraints on the reading of functors in formal languages into English; but one should be wary of drawing exaggerated metaphysical or conceptual implications from the rules of English grammar. (We intend this thesis to be interpreted in such a way that it is consistent with the view that indeed *some* features of English grammar–e.g., its love for the subject-predicate form–can be argued to be metaphysically significant.)

It is a crucial part of the argument that, on our view, the propositional predicate "— is true" and the propositional subnector "that —" can be viewed as sometimes (we by no means say always) playing low grade and philosophically uninteresting roles in English: that of converting a term into a sentence, and that of converting a sentence into a term, respectively, without semantic increment. (Of course the thesis that such and such is of no philosophical interest can itself be interesting and debatable.) The idea is that English has certain constructions requiring terms and others requring sentences; so if one has somehow started with one but English requires the other, one simply goes to the shelf and pulls out a "that —" or a "— is true." Examples go both ways, as we have illustrated above: to negate a complex sentence appears to require prior nominalization with "that —," while to propositionally conjoin seems to require prior sententialization with "— is true." And, we argue, that a construction in English needs in one case a term and in another a sentence and therefore calls for prenominalizaton or presententialization is to be viewed as a mere, sheer accident, unless independent philosophical argument can show otherwise.

Our use of the word "accident" needs clarification. Linguists divide linguistic features into those that are present in only some languages and those that are true linguistic universals (if there are any). The non-universal features might be called "accidental," since not lawlike, since not universal; but the question as to whether or not a feature is of philosophic importance is wholly independent of the question of universality. For example, the subject-predicate form may well be of profound conceptual interest even if non-universal; and contrariwise, even if biological constraints should lead to a law that certain types of self-embedding constructions can never, in any language, nest more deeply than (say) depth 7, that fact, though not of course devoid of interest for certain parts of philosophy, would not have the kind of conceptual implications which would render it of interest to the logical grammarian. It would be a kind of *conceptual* accident even though an *empirical* law; and it is this sense of "accident" we have in mind. The upshot is that one can never take a feature, whether universal or not, as of philosophical import without some kind of further and distinctively philosophical argument.

Although Hiż, as a linguist, may not wish to be associated with a view which reads certain linguistic features as "accidents," we quote Hiż [5] in support of the innocuousness of subnectors: "Nominalized forms of sentences are not names of the sentences. Rather they are different forms of the very same sentences." A corollary is that one should not feel one is "committed" to propositions because he is dealing in constructions like negation, requiring prenominalization with "that —," nor should he feel he must be "committed" to a notion of propositional truth or to some kind of primacy of sentences because constructions like propositional conjunction require presententialization with "— is true." Such "commitments" would arise only from further and less innocuous uses of "that" and "true." So:

THESIS 8: One of the roles of each of "that" and "is true" in English is to serve as a grammatical dodge to evade the constraints of English grammar without semantic penalty.

So much for theses.

A4. A Word about Quantifiers

It might be thought that against our downplaying of the term-sentence distinction one should count the quantification structure of English, which leads us to ontological commitment and all that by means of its permission to generalize with respect to term positions, while forbidding generalization with respect to sentence positions. For example, there are the well-known difficulties concerning reading propositional quantifiers into English. We think, however, that Grover [4] shows conclusively that these difficulties are due to "accidental" features of English grammar and accordingly both illustrates and supports our general thesis.

A5. Conditional and Entailment

Let us, finally, proceed to a consideration of conditionals and entailments from the point of view of logical grammar. An analogue to Pilate's question about truth is: What is Entailment? We follow for a while the numbering of our theses.

1. "— entails —" is a two-place predicate, taking sentence- or proposition-terms as input and producing a sentence as output; and, of course, "if — then —" is a connective.

2. Without disputing the usefulness of using "conditional" to refer to English constructions based on "if," we recognize the existence of a generic conditional-entailment-implication concept.

3. The table shows that the simplest functors in the vicinity of this generic concept are the "if — then —" connective and the "— entails —" predicate.

4a. From this it by no means follows that entailment is a relation; to say so would constitute a naive and slavish adherence to the idiosyncracies of English grammar, for entailment could also be represented by any of an operator, a connective, or a subnector, provided we allow enough presententializing and prenominalizing.

4b. The only or easiest way to carry entailment as a connective involves prenominalizing by "that"; which is to say, the entailment connective in English is to be taken as "that — entails that —."

We remark further that the English entailment connective, like any connective, takes sentences as both input and output, and therefore nests in English without limit. Can we responsibly avoid investigating the complexities deriving therefrom?

5. English admits "entails" as both a proposition predicate and a sentence predicate; grammar certainly gives us no reason to doubt that the theories of these two should be equally interesting.

6. Although "entails" in English is unmistakably a predicate, this grammatical fact should not lead us to forego investigation of entailment qua connective. Indeed, although it does not represent one of our principal interests, it is symmetrically in order to remark that the if-then predicate, "if — is true, then — is true" provides an object of legitimate study. (If the underlying if-then were truth functional, we should be in the presence of the relation of material "implication.")

7. In [1] we use the arrow, so that, using "A" and "B" as sentences, "A→B" is a sentence. Accordingly, when careful we must read a formula "A→B" into English as "that A entails that B." And when we raise the question of the applicability to English of the logic of entailment, we intend the application to be made via the "that — entails that —" connective. (But see below concerning the interchangeability of "if-then" and "entails.")

8. We do not think the instances of "that" in "that — entails that —" commit us to propositions. We think it an accident that in English entailment is carried by a functor requiring terms instead of sentences and thus requires prenominalization. We take the immediate constituents of "that Tom runs entails that Tom moves" to be "Tom runs," "Tom moves," and the above exhibited connective, with the occurrences of "that" playing an only grammatical role. (We like propositions all right; it is just that we don't want to be "committed" to them before we are ready; we want to be committed by our theory, and not by our grammar. We take ourselves in this respect to be in parallel with the fellow using "it is not the case that —.")

In accordance with 8, while in partial opposition to 7, it must be said that we do not much worry about dropping an occasional "that" from time to time. We feel "snow is white does not entail that snow is white entails itself" is more readable than the more correct "that snow is white does not entail that that snow is white entails itself," and is wholly unambiguous. If we thought of "that" in this sort of example as of philosophical interest, we'd worry; but we don't. Similarly, even when the letters are intended to take the places of sentences, we may sometimes be found reading "A→B" as "A entails B," instead of "that A entails that B." We hope this grammatical propaedeutic will enable any reader of [1] to sprinkle "that" around enough to save us from solecism.

We have now run out of theses, but we find ourselves with two more important points to make. The first has to do with our endemic tendency to read "A→B" as "if A then B" as often as we read it with "entails." In justification of this procedure, we argue that the most significant difference between "if-then" and "entails" (or "implies") is that the first functor is a connective—calls for sentences—whereas the latter is a predicate—calls for terms. Repeat: the *most* significant difference. And if we are right, it is not much of a difference at all except to a grammarian. We think nothing philosophical hangs on this famous distinction. It is a mere, sheer, clear fact of grammar.

We further think there are all kinds of conditionals (uses of "if-then") in English, and also all kinds of uses of the entailment or implication connective; and that many of them correspond in a one-to-one fashion. E.g., "that A logically entails that B" has "if A then, as a matter of logic, B," with ellipsis in both cases producing respectively "that A entails that B" and "if A then B." In particular, we think every use of "implies" or "entails" as a connective can be replaced by a suitable "if-then"; however, the converse may not be true. But with reference to the uses in which we are primarily interested, we feel free to move back and forth between "if-then" and "entails" in a freewheeling manner.

Our very last point is a little phony, since it presupposes having to make a decision as to whether to treat entailment as a sentence predicate on the one hand or as a connective on the other; and after all, Peirce's warning not to block the road to inquiry should lead us to embrace both alternatives as legitimate. But suppose we have to make a decision; then the final point we want to make is that there are costs on both sides. The cost of taking entailment as a connective is often remarked and well known: we must pay attention to the nesting of entailments within entailments, and so on without end. Or perhaps "must" is a bit strong here, but in any event the grammatical fact that connectives use sentences as both input and output strongly suggests attention to the problem of nesting. In contrast, if we take entailment as a predicate, the grammar forbids nesting; for the output of a predicate is a sentence, whereas its inputs must be terms. Again this is somewhat of an oversimplification, since we might wish to deal with entailment sentences whose terms name other entailment sentences; but ignoring that, the theory of the entailment predicate will be simpler than the theory of the entailment connective by avoiding nesting. But the cost is by no means all on one side, a point that is seldom noticed. For to take entailment as a sentence predicate means that one has to involve himself with all the bag and baggage of The Metalanguage, which as we know very well is not only dreadfully complicated in its syntactic department, as anyone who has set out to completely formalize these matters can testify, but also with respect to its semantic ingredients skirts the very edge of paradox, so that one taking this road must exercise not only care but caution. So let no one suppose that departure from simplicity is a peculiar characteristic of the theory of the entailment connective.

We summarize the spirit of this propaedeutic with a bit of advice meant to parallel some most of us received in the earlier stages of our education: learn logical grammar thoroughly, and then—but only then—be relaxed about it.

References

[1] Anderson, A. R. and Belnap, N. D. Jr. 1975. *Entailment: the logic of relevance and necessity.* Princeton: Princeton University Press.

[2] Curry, H. B. 1963. *Foundations of mathematical logic.* New York: McGraw-Hill.

[3] Curry, H. B. and Feys, R. 1958. *Combinatory logic, vol. 1.* Amsterdam: North Holland.

[4] Grover, D. L. 1972. Propositional quantifiers. *Journal of Philosophical Logic* 1: 111–36.

[5] Hiż, H. 1961. Modalities and extended systems. *The Journal of Philosophy* 68: 723–31.

11: RICHMOND H. THOMASON

Decidability in the Logic of Conditionals*

In Stalnaker and Thomason [12] an axiomatic system **CQ** of first-order modal logic was proposed as a theory sufficiently general to capture subjunctive as well as indicative uses of the conditional in English. Most of [12] is devoted to a proof that **CQ** is sound and complete relative to a semantic interpretaton using intensional model structures.[1] The existence of a completeness proof, of course, does not in any way justify the claim that **CQ** provides a semantic theory of the conditional as it is used in English. Though there is much evidence that is relevant ot this claim (especially, evidence concerning the truth values of conditional sentences in English), the status and interpretation of this evidence remains controversial.[2] Arguments attempting to show that this evidence weighs in favor of a particular theory, such as that embodied in **CQ**, can carry a high degree of plausibility, but such justifications have not as yet proved convincing enough to put an end to controversy.

In [11], Stalnaker seeks ot justify the semantics of **CQ**, referring to the philosophical literature on subjunctive conditionals and to general intuitions concerning the meaning of conditional sentences, as well as to the truth values of particular sentences. In Thomason [13], I attempt to produce a justification of the proof theory of the system. To do this I make use of the framework of subordinate proofs developed by Fitch in [3], first reformulating the sentential fragment of **CQ** within this framework. Theories of natural deduction are in general superior to axiomatic theories in their affinity to the way people actually reason, but even when compared with other natural deduction formats Fitch's is remarkable for its "naturalness" in this respect. Exploiting this feature, I try to show

*This paper owes much to conversations with Robert Stalnaker. In particular, the definition of the ordering characterized in *D4.2* should be attributed to him. An earlier draft of this paper was written under the support of National Science Foundation Grant GS–2517.

in [13] that **CQ** exemplifies patterns of argumentation that in fact occur typically in specimens of reasoning with conditional sentences.

Like [12], the present paper is concerned only with **CS**, the sentential fragment of **CQ**. It takes up the purely technical problem of showing **CS** to be decidable. I will use a semantic type of proof which, as a dividend, exemplifies a highly general method of showing that sentential calculi have the finite model property and so are decidable. Though, of course, a number of methods of proving decidability—especially in modal logic—have been discussed in the literature,[3] this method is not, as far as I know, one of them.

2. *Preliminary Definitions*

This paper presupposes the material on conditional logic mentioned above, especially that presented in [12]. To render it more self-contained, however, I will present here the basic proof-theoretic and semantic notions pertaining to **CS**, the sentential fragment of **CQ**. Definitions *D2.1* to *D2.9* parallel ones to be found in [12].

D2.1. The axioms of **CS** are determined by AO and axiom schemes A1–A6; its only primitive rules of inference are modus ponens and necessitation (from A to infer $\Box A$). Necessity is defined in terms of the conditional connective $>$: $A =_{df} \sim A > \sim(P \supset P)$, where P is a fixed sentence parameter. Similarly, $\Diamond A =_{df} \sim(A > \sim(P \supset P))$.[4]

A0. Any formula that is a substitution instance of a tautology is an axiom.

A1. $\Box(A \supset B) \supset (\Box A \supset \Box B)$

A2. $\Box(A \supset B) \supset (A > B)$

A3. $\Diamond A \supset ((A > B) \supset \sim(A > \sim B))$

A4. $(A > (B \vee C)) \supset ((A > B) \vee (A > C))$

A5. $(A > B) \supset (A \supset B)$

A6. $((A > B) \wedge (B > A)) \supset ((A > C) \supset (B > C))$

D2.2. A *model structure* is a triple $\langle \mathcal{K}, \mathcal{R}, \lambda \rangle$, where $\mathcal{K}$ is a non-empty set, $\lambda \in \mathcal{K}$, and $\mathcal{R}$ is a binary, reflexive relation on $\mathcal{K}'$, where $\mathcal{K}' = \mathcal{K} - \{\lambda\}$.

D2.3. A *morphology* is a nonempty set **M**. The members of **M** are called the *atomic sentences,* or *sentence parameters,* of **M**. The set $F_{\mathbf{M}}$ of formulas of **M** is the smallest set Z such that $\mathbf{M} \subseteq Z$ and $\{\sim A, A \supset B, A > B\} \subseteq Z$ for all $A,B \in Z$.

D2.4. A *valuation* of a mortpholgy **M** on a model structure $\langle \mathscr{K}, \mathscr{R}, \lambda \rangle$ is a function v from $\mathscr{K}' \times \mathbf{M}$ to $\{T,F\}$.

D2.5. An *s-function* on a model structure $\langle \mathscr{K}, \mathscr{R}, \lambda \rangle$ and morphology **M** is a function f from $F_{\mathbf{M}} \times \mathscr{K}'$ to $\mathscr{K}$, such that for all formulas A of **M** and $\alpha \in \mathscr{K}'$, if $f(A,\alpha) \neq \lambda$ then $\alpha \mathscr{R} f(A,\alpha)$.

D2.6. A *quasi-interpretation* I of a morphology **M** on a model structure $\langle \mathscr{K}, \mathscr{R}, \lambda \rangle$ is a pair $\langle v,f \rangle$, where v is a valuation of **M** on $\langle \mathscr{K}, \mathscr{R}, \lambda \rangle$ and f is an s-function on $\langle \mathscr{K}, \mathscr{R}, \lambda \rangle$ and **M**.

D2.7. Let $I = \langle v,f \rangle$ be a quasi-interpretation of **M** on $\langle \mathscr{K}, \mathscr{R}, \lambda \rangle$. Then $I_\lambda(A) = T$ for all $A \in F_{\mathbf{M}}$, and for all $\alpha \in \mathscr{K}'$:

(1) $I_\alpha(P) = v(P,\alpha)$, where $P \in \mathbf{M}$;

(2) $I_\alpha(\sim A) = T$ iff $I_\alpha(A) = F$;

(3) $I_\alpha(A \supset B) = T$ iff $I_\alpha(A) = F$ or $I_\alpha(B) = T$;

(4) $I_\alpha(A > B) = T$ iff $I_{f(A,\alpha)}(B) = T$.

D2.8. An *interpretation* of **M** on a model structure $\langle \mathscr{K}, \mathscr{R}, \lambda \rangle$ is a quasi-interpretation $I = \langle v,f \rangle$ of **M** on $\langle \mathscr{K}, \mathscr{R}, \lambda \rangle$ such that, for all $A,B \in F_{\mathbf{M}}$ and $\alpha \in \mathscr{K}'$:

(1) $I_{f(A,\alpha)}(A) = T$;

(2) if $I_\alpha(A) = T$ then $f(A,\alpha) = \alpha$;

(3) if $f(A,\alpha) = \lambda$ then there is no $\beta \in \mathscr{K}'$ such that $\alpha \mathscr{R} \beta$ and $I_\beta(A) = T$;

(4) if $I_{f(A,\alpha)}(B) = I_{f(B,\alpha)}(A) = T$ then $f(A,\alpha) = f(B,\alpha)$.

D2.9. A set Γ of formulas of **M** is *simultaneously satisfiable* on a model structure $\mathscr{M} = \langle \mathscr{K}, \mathscr{R}, \lambda \rangle$ if there is an interpretation I of **M** on $\mathscr{M}$ and an $\alpha \in \mathscr{K}'$ such that $I_\alpha(A) = T$ for all $\alpha \in \mathscr{K}'$.

3. Partial Interpretations

The following definitions will be needed in the proof that **CS** is decidable.

D3.1. The *rank* r(A) of a formula A of **M** is defined inductively: if $A \in \mathbf{M}$ then r(A) = 0; r($A \supset B$) = max(r(A), r(B)), r($\sim A$) = r(A), and r($A > B$) = 1 + max(r(A), r(B)).

D3.2. Where **M** is a morphology, $F^k_{\mathbf{M}}$ is the set of formulas of **M** having rank not exceeding k.

D3.3. A *partial s-function* on a model structure $\langle \mathscr{K}, \mathscr{R}, \lambda \rangle$ and morphology **M** is a partial function f from $F_{\mathbf{M}} \times \mathscr{K}'$ to $\mathscr{K}$, such that if f(A,α) $\neq \lambda$ then $\alpha \mathscr{R}$ f(A,α). In this paper I will confine my attention to *regular* partial s-functions, which have two additional properties: (1) for all $\alpha, \beta \in \mathscr{K}'$, if $\alpha \mathscr{R} \beta$, then there is an $A \in F_{\mathbf{M}}$ such that f(A,α) = β, and (2) for all $\alpha \in \mathscr{K}'$ there is a nonnegative integer k_α such that for all $A \in F_{\mathbf{M}}$, f(A,α) is defined iff r(A) $< k_\alpha$.

The most important thing about a regular partial s-function is that whether or not it is defined for a formula A at a given α will depend only on the rank of A. Below, whenever I speak of "partial s-functions" I take them to be regular. Note, incidentally, that no regular partial s-function can be total.

D3.4. A *partial quasi-interpretation* I of a morphology **M** on a model structure $\mathscr{M} = \langle \mathscr{K}, \mathscr{R}, \lambda \rangle$ is a pair $\langle$v,f$\rangle$ such that v is a valuation of **M** on $\mathscr{M}$ and f is a partial s-function on $\mathscr{M}$ and **M.** By means of an inductive definition, partial quasi-interpretations can be made to assign truth values to formulas. The *satisfaction function* H associated with a partial quasi-interpretation I is the smallest partial function G from $\mathscr{K}' \times F_{\mathbf{M}}$ to $\{T,F\}$ such that v $\subseteq$ G and G meets the conditions of *D2.7,* supplemented by appropriate conditions for the falsity of formulas: e.g., H(α,$\sim A$) = F iff H(α,A) = T.[5] Where H is the satisfaction function associated with I, $I_\alpha(A)$ is H(α,A).

D3.5. A partial interpretation of a morphology **M** on a model structure $\mathscr{M} = \langle \mathscr{K}, \mathscr{R}, \lambda \rangle$ is a partial quasi-interpretation of **M** on $\mathscr{M}$, such that for all $A,B \in F_{\mathbf{M}}$ and $\alpha \in \mathscr{K}'$:

(1) $I_{f(A,\alpha)}(A)$ = T if f(A,α) is defined;

(2) if $I_\alpha(A)$ = T and f(A,α) is defined then f(A,α) = α;

(3) if f(A,α) = λ then $I_\beta(A)$ = F for all $\beta \in \mathscr{K}'$ such that $\alpha \mathscr{R} \beta$;

(4) if f(A,α) and f(B,α) are defined and $I_{f(A,\alpha)}(B) = I_{f(B,\alpha)}(A)$ = T, then f(A,α) = f(B,α).

D3.6. A partial interpretation I = ⟨v,f⟩ of **M** on $\langle \mathscr{K}, \mathscr{R}, \lambda \rangle$ is *regular* if f is regular and furthermore, for all $A \in F_{\mathbf{M}}$ and $\alpha,\beta \in \mathscr{K}'$, if $f(A,\alpha) = \beta$ then $I_\beta(A)$ is defined.

Whenever I speak of "partial interpretations" below, they are understood to be regular.

The following lemma is stated without proof; it follows immediately from *D3.3* and *D3.5*.

L3.1. Let I be a partial interpretation of a morphology **M** on a model structure $\langle \mathscr{K}, \mathscr{R}, \lambda \rangle$. Let $\alpha \in \mathscr{K}'$, and let k_α be as in *D3.3*. Then $I_\alpha(A)$ is defined iff $A \in F^{k}\alpha$.

4. *Extending Partial Interpretations*

The definitions and lemmas in this section are means to showing that every partial interpretation of **M** on $\langle \mathscr{K}, \mathscr{R}, \lambda \rangle$ can be extended to an interpretation of **M** on $\langle \mathscr{K}, \mathscr{R}, \lambda \rangle$, provided that $\mathscr{K}$ is finite.

D4.1. Let I = ⟨v,f⟩ be a partial interpretation of **M** on $\mathscr{M} = \langle \mathscr{K}, \mathscr{R}, \lambda \rangle$. Then $d(I,\mathscr{M})$ is the smallest number j such that for some $\alpha \in \mathscr{K}'$, $k_\alpha = j$.

D4.2. Let I and $\mathscr{M}$ be as in *D4.1*, and let $\alpha \in \mathscr{K}'$, and $\alpha \mathscr{R} \beta$, $\alpha \mathscr{R} \gamma$. Then $\beta \leqslant_\alpha \gamma$ if there exists an $A \in F_{\mathbf{M}}$ such that $f(A,\alpha) = \beta$ and $I_\gamma(A) = T$. Furthermore, $\beta \leqslant_\alpha \lambda$ if $\alpha \mathscr{R} \beta$ or $\beta = \lambda$.

L4.1. Let I and $\mathscr{M}$ be as in *D4.1*, and let $\alpha \in \mathscr{K}'$. Then if $f(A \vee \beta, \alpha)$ is defined, $f(A \vee B,\alpha) = f(A,\alpha)$ or $f(A \vee \beta,\alpha) = f(\beta,\alpha)$.
Proof. Since I is a partial interpretation and $f(A \vee B,\alpha)$ is defined, $I_{f(A \vee B,\alpha)}(A \vee B) = T$, so either (1) $I_{f(A \vee B,\alpha)}(A) = T$ or (2) $I_{f(A \vee B,\alpha)}(B) = T$. In case 1, $I_{f(A \vee B,\alpha)}(A) = I_{f(A,\alpha)}(A \vee B) = T$, and therefore $f(A \vee B,\alpha) = f(A,\alpha)$. In case 2, $I_{f(A \vee B,\alpha)}(B) = I_{f(\beta,\alpha)}(A \vee B) = T$, and therefore $f(A \vee \beta,\alpha) = f(B,\alpha)$.

L4.2. Let I and $\mathscr{M}$ be as in *D4.1*, let $\alpha \in \mathscr{K}'$, and let $\beta,\gamma \in \{\delta / \alpha \mathscr{R} \beta$ or $\delta = \lambda\}$. Then $\beta = \gamma$ if $\beta \leqslant_\alpha \gamma$ and $\gamma \leqslant_\alpha \beta$.
Proof. If $\beta \leqslant_\alpha \gamma$ and $\gamma \leqslant_\alpha \beta$ then for some $A,B \in F_{\mathbf{M}}$, $f(A,\alpha) = \beta$ and $I_\gamma(A) = T$, and $f(B,\alpha) = \gamma$ and $I_\beta(B) = T$. But then $I_{f(B,\alpha)}(A) = I_{f(A,\alpha)}(B) = T$, so $\beta = \gamma$.

L4.3. Let I and $\mathscr{M}$ be as in *D4.1*, let $\alpha \in \mathscr{K}'$, and let $\beta,\gamma \in \{\delta / \alpha \mathscr{R} \delta$ or $\delta = \lambda\}$. Then either $\beta \leqslant_\alpha \gamma$ or $\gamma \leqslant_\alpha \beta$.
Proof. Let $\beta = f(A,\alpha)$ and $\gamma = f(B,\alpha)$. By *L4.1*, either $f(A \vee B,\alpha) = \beta$ or $f(A \vee B,\alpha) = \gamma$. Now, $I_\beta(A \vee B) = I_\gamma(A \vee B) = T$. Then

$\beta \leqslant_\alpha \gamma$ if $f(A \vee B,\alpha) = \beta$ and $\gamma \leqslant_\alpha \beta$ if $f(A \vee B,\beta) = \gamma$. It follows that either $\beta \leqslant_\alpha \gamma$ or $\gamma \leqslant_\alpha \beta$.

L4.4. Let I and $\mathscr{M}$ be as in *D4.1*, let $\alpha \in \mathscr{K}'$, and let $\beta,\gamma,\delta \in \{\zeta/\alpha\mathscr{R}\zeta$ or $\zeta = \lambda\}$. Then $\beta \leqslant_\alpha \delta$ if $\beta \leqslant_\alpha \gamma$ and $\gamma \leqslant_\alpha \delta$.
Proof. If $\beta \leqslant_\alpha \gamma$ and $\gamma \leqslant_\alpha \delta$ then for some $A,B \in F_{\mathbf{M}}$, $f(A,\alpha) = \beta$ and $I_\gamma\ (A) = T$, and $f(B,\alpha) = \gamma$ and $I_\delta\ (B) = T$. By *L4.1*, $f(A \vee B,\alpha) = \beta$ or $f(A \vee B,\alpha) = \gamma$. If $f(A \vee B,\alpha) = \gamma$ then $I_{f(A \vee B,\alpha)}\ (A) = I_{f(A,\alpha)}\ (A \vee B) = T$ so that $\beta = \gamma$ and therefore $\beta \leqslant_\alpha \delta$. If $f(A \vee B,\alpha) = \beta$ then, since $I_\delta\ (A \vee B) = T$, $\beta \leqslant_\alpha \delta$. In either case, $\beta \leqslant_\alpha \delta$.

L4.5. Let $I = \langle v,f \rangle$ be a partial interpretation of **M** on a finite model structure $\mathscr{M} = \langle \mathscr{K}, \mathscr{R}, \lambda \rangle$ (this means that the set $\mathscr{K}$ is finite). Then for all $\alpha \in \mathscr{K}'$, the set $\{\beta/\alpha\mathscr{R}\beta$ or $\beta = \lambda\}$ under the ordering $\leqslant_\alpha$ is isomorphic to some initial set $\{i/i < j\}$ of nonnegative integers under the ordering $\leqslant$. Moreover, the least member of $\{\beta/\alpha\mathscr{R}\beta$ or $\beta = \lambda\}$ under $\leqslant_\alpha$ is α, and the greatest member is λ.
Proof. The set $\{\beta/\alpha\mathscr{R}\beta$ or $\beta = \lambda\}$ must be finite by assumption. But any finite set ordered by a reflexive relation satisfying the conditions established in *L4.2, L4.3,* and *L4.4* is isomorphic to an initial set of integers under the ordering $\leqslant$. Also $\alpha \leqslant_\alpha \beta$ for all $\beta \in \{\gamma/\alpha\mathscr{R}\gamma$ or $\gamma = \lambda\}$ because if $I_\alpha\ (A) = T$ then $f(A,\alpha) = \alpha$; and $\beta \leqslant_\alpha \lambda$ for all $\beta \in \{\gamma/\alpha\mathscr{R}\gamma$ or $\gamma = \lambda\}$ by *D4.2*.

L4.6. Let v be a valuation of **M** on $\mathscr{M}$, and let f and f′ be partial s-functions on $\mathscr{M}$ such that f′ is an extension of f. Let $I = \langle v,f \rangle$ and $I' = \langle v,f' \rangle$. Then for all $\alpha \in \mathscr{K}'$ and all $A \in F_{\mathbf{M}}$ such that $I_\alpha(A)$ is defined, $I_\alpha(A) = I'_\alpha(A)$.
Proof. This is easily shown by induction on the complexity of *A*.

D4.3. Let I and $\mathscr{M}$ be as in *L4.5*, let $\alpha \in \mathscr{K}'$ and let *A* be a formula of **M** such that $I_\beta\ (A)$ is defined for all β such that $\alpha\mathscr{R}\beta$. Then $m(I,A,\alpha)$ is the least member β of $\{\gamma/\alpha\mathscr{R}\gamma$ or $\gamma = \lambda\}$, with respect to the ordering $\leqslant_\alpha$, such that $I_\beta\ (A) = T$. If $I_\beta\ (A)$ is undefined for some β such that $\alpha\mathscr{R}\beta$, then $m(I,A,\alpha)$ is also undefined.

L4.7. Let I, $\mathscr{M}$, α, and *A* be as in *D4.3.* Then if $f(A,\alpha)$ is defined, $f(A,\alpha) = m(I,A,\alpha)$.
Proof. Since I is a partial interpretation, $I_{f(A,\alpha)}\ (A) = T$. Therefore, in view of *D4.3*, $m(I,A,\alpha) \leqslant_\alpha f(A,\alpha)$. But since $f_{m(I,A,\alpha)}\ (A) = T$, it follows from *D4.2* that $f(A,\alpha) \leqslant_\alpha m(I,A,\alpha)$. By *L4.2*, $f(A,\alpha) = m(I,A,\alpha)$.

D4.4. Let I and $\mathscr{M}$ be as in *L4.5,* and let $f^1(A,\alpha) = m(I,A,\alpha)$ for all A such that $m(I,A,\alpha)$ is defined. Also let I^1 be $\langle v,f^1 \rangle$.

L4.8. Let I, $\mathscr{M}$, α, and A be as in *D4.3.* If $m(I,A,\alpha) \neq \lambda$ then $\alpha \mathscr{R} m(I,A,\alpha)$.
Proof. This is immediate from *D4.3.*

L4.9. Let I, $\mathscr{M}$, α, and A be as in *D4.3.* Then $I_{m(I,A,\alpha)}(A) = T$.
Proof. This too is immediate from *D4.3.*

L4.10. Let I and $\mathscr{M}$ be as in *L4.5.* Then I^1 is a partial interpretation of **M** on $\mathscr{M}$ and I^1 is an extension of I, in the sense that for all $A \in F_{\mathbf{M}}$ and $\alpha \in \mathscr{K}'$, if $I_\alpha(A)$ is defined then $I_\alpha(A) = I^1_\alpha(A)$. Furthermore, $d(I,\mathscr{M}) < d(I^1,\mathscr{M})$.
Proof. *L4.8* shows f^1 to be a partial s-function on $\mathscr{M}$ and **M.** (Note that f^1 is regular because, in view of *L3.1,* $f^1(\alpha,A)$ is defined iff $A \in F^j$, where j is the smallest n such that for some β such that $\alpha \mathscr{R} \beta$, $j = k_\beta$.) Thus, I^1 is a partial quasi-interpretation of $\mathscr{M}$ on **M**.
Now consider conditions 1–4 of *D3.14.* As for condition 1, $I^1_{f^1(A,\alpha)}(A) = T$ if $f^1(A,\alpha)$ is defined, by *L4.9* and *L4.6.* As for condition 2, suppose that $I^1_\alpha(A) = T$ and $f^1(A,\alpha)$ is defined. Then, since $f^1(A,\alpha)$ is defined, $I_\beta(A)$ is defined for all β such that $\alpha \mathscr{R} \beta$, and in particular $I_\alpha(A)$ is defined. Then, by *L4.6,* $I_\alpha(A) = T$. Since by *L4.3* α is the least member of $\{\beta/\alpha \mathscr{R} \beta \text{ or } \beta = \lambda\}$ under $\leqslant_\alpha$, $f^1(A,\alpha) = \alpha$ by *D4.3.* As for condition 3, suppose $f^1(A,\alpha) = \lambda$. Since $f^1(A,\alpha)$ is defined, $I_\beta(A)$ is defined for all β such that $\alpha \mathscr{R} \beta$. If $I_\beta(A) = T$ for some β such that $\alpha \mathscr{R} \beta$, then $\lambda \leqslant_\alpha \beta$, contradicting *L4.5.* Hence $I_\beta(A) = F$ for all β such that $\alpha \mathscr{R} \beta$. As for condition 4, suppose that $f^1(A,\alpha) = \beta$ and $f^1(B,\alpha) = \gamma$, and that $I^1_\beta(A) = I^1_\gamma(B) = T$. Then, since $\alpha \mathscr{R} \beta$ and $\alpha \mathscr{R} \gamma$, $I_\beta(A)$ and $I_\gamma(B)$ are defined, and by *L4.10* $I_\beta(A) = I_\gamma(B) = T$. Since I is a partial interpretation, $\beta = \gamma$.
This shows I^1 to be a partial interpretation of **M** on $\mathscr{M}$. By *L4.7,* f^1 is an extension of f; it follows from *L4.6* that I^1 is an extension of I. Furthermore, suppose that $d(I,\mathscr{M}) = j$ and let $\alpha \in \mathscr{K}'$. By *L3.1,* for all $\beta \in \mathscr{K}'$ such that $\alpha \mathscr{R} \beta$, $I_\beta(A)$ is defined for all $A \in F^{k_\beta}_{\mathbf{M}}$, and so $f^1(A,\alpha)$ is defined for all A such that $r(A) < k_\beta + 1$. Therefore $d(I^1,\mathscr{M}) = d(I,\mathscr{M}) + 1$.

T4.1. Let I be a partial interpretation of **M** on $\mathscr{M}$, where $\mathscr{M}$ is finite. Then I is extendable to an interpretation I^* of **M** on $\mathscr{M}$.
Proof. Let I^{n+1} be $(I^n)^1$, and let I^* be the limit of the sequence I, $I^1, I^2, \ldots$ of partial interpretations. By *L4.10,* $d(I^{n+1},\mathscr{M}) >$

$d(I^n, \mathcal{M})$ for all n; hence $f^*(A,\alpha)$ is defined for all $A \in F_M$ and $\alpha \in \mathcal{K}$, and so I^* is a quasi-interpretation. By *L4.10,* I^n is a partial interpretation for each n; it follows immediately that I^* is a interpretation.

T4.2. Let Γ be a set of formulas of **M**. If there is a partial interpretation I of **M** on a finite model structure $\mathcal{M} = \langle \mathcal{K}, \mathcal{R}, \lambda \rangle$ such that $I_\alpha(A) = T$ for all $A \in \Gamma$ at some fixed $\alpha \in \mathcal{K}'$, then Γ is simultaneously satisfiable on $\mathcal{M}$.
Proof. Under the conditions of the theorem, *T4.1* guarantees that there is an interpretation I^* of **M** on $\mathcal{M}$ that is an extension of I. Since $I_\alpha(A) = T$ for all $A \in \Gamma$, $I^*_\alpha(A) = T$ for all $A \in \Gamma$.

5. Semantic Completeness, Finite Models, and Decidability

In [12] it is shown that if a set Γ of formulas is consistent then Γ is simultaneously satisfiable.[6] Here this proof will be refined to show that if a finite set Γ of formulas is consistent then there is a partial interpretation on a finite model structure $\mathcal{M} = \langle \mathcal{K}, \mathcal{R}, \lambda \rangle$ and an $\alpha \in \mathcal{K}'$ such that $I_\alpha(A) = T$ for all $A \in \Gamma$. From this and *T4.2* the finite model property follows immediately.

D5.1. A set Γ of formulas is $\mathbf{M}^k$*-saturated* if (1) $\Gamma \subseteq F^k_M$, (2) Γ is consistent, and (3) for all $A \in F^k_M$, $A \in \Gamma$ or $\sim A \in \Gamma$.

D5.2. Where Γ is $\mathbf{M}^k$*-saturated* and $k > 0$, let $h(\Gamma) = \{\{B/A > B \in \Gamma\} / A > B \in F^k_M \text{ and } \Diamond A \in \Gamma\}$. If $k = 0$, $h(\Gamma)$ is undefined. If S is a set of $\mathbf{M}^k$-saturated sets, $h(S) = \{h(\Gamma) / \Gamma \in S\}$. Note that if Γ is $\mathbf{M}^0$-saturated for all $\Gamma \in S$, $h(\Gamma) = \emptyset$.

L5.1. If Γ is $\mathbf{M}^k$-saturated, then for all $\Delta \in h(\Gamma)$, there is a $j < k$ such that Δ is $\mathbf{M}^j$-saturated.
Proof. Compare the proof of *L5.5* in [12]

L5.2. If **M** is finite then there exist only finitely many formulas of F^k_M, up to provable equivalence.
Proof. If **M** has cardinality n then there are $2^{(2^n)}$ formulas of **M**, up to provable equivalence. Since if $\vdash A \equiv B$ and $\vdash C \equiv D$, then $\vdash (A > C) \equiv (B > D)$, it follows that for any k, there will be only finitely many formulas of $\mathbf{M}^k$, up to provable equivalence.

L5.3. If $\mathbf{M}^k$ is finite and Γ is $\mathbf{M}^k$-saturated then $h(\Gamma)$ is finite.
Proof. This follows from *L5.2,* together with the fact that if $\vdash A \equiv B$, then $\{C / A > B \in \Gamma\} = \{C / B > C \in \Gamma\}$.

L5.4. Let $\mathscr{K}'(\Gamma) = \{\Gamma\} \cup h(\{\Gamma\}) \cup hh(\{\Gamma\}) \cup \ldots$, where Γ is $\mathbf{M}^k$-saturated. If **M** is finite then $\mathscr{K}'(\Gamma)$ is finite.
Proof. Look at $\mathscr{K}'(\Gamma)$ as a tree generated from the node Γ by the operation h. *L5.3* shows that each of its nodes has finitely many nodes immediately above it. On the other hand, each chain through the tree has finite height, since by *L5.1* each step upwards in the tree reduces rank of saturation, and if Δ is $\mathbf{M}^0$-saturated then $h(\Delta)$ is undefined. It follows by König's lemma that $\mathscr{K}'(\Gamma)$ is finite.

L5.5. Let **M** be finite and Γ be $\mathbf{M}^k$-saturated. Then there is a partial interpretation $I = \langle v,f \rangle$ of **M** on a model structure $\langle \mathscr{K}, \mathscr{R}, \lambda \rangle$ with $\mathscr{K}$ finite such that for some $\alpha \in \mathscr{K}$, $\Gamma = \{A / I_\alpha(A) = T\}$.
Proof. The set $\mathscr{K}'$ is $\mathscr{K}'(\Gamma)$; λ is $F^k_{\mathbf{M}}$. The relation $\mathscr{R}$ is defined by letting $\Delta \mathscr{R} \Theta$ iff $\Theta \in h(\Delta)$. The partial interpretation I is defined as usual in Henkin-style proofs (in particular, see *D6.5* and *D6.6* of [12]. For example, $f(A,\Delta) = \{B / A > B \in \Delta\}$ for A such that $r(A) < j$, where Δ is $\mathbf{M}^j$-saturated. Thus, $f(A,\Delta)$ will be $\mathbf{M}^{j-(r(A)+1)}$-saturated if $j > r(A)$ and undefined if $j \leqslant r(A)$.

L5.6. Let **M** be any morphology. If Γ is a consistent subset of $F^k_{\mathbf{M}}$ then Γ has an $\mathbf{M}^k$-saturated extension.
Proof. This lemma is established in the usual way by adding formulas of $\mathbf{M}^k$ to Γ in a consistent manner, and then taking the limit of this process.

L5.7. Let **M** be a finite morphology. If Γ is a consistent subset of $F^k_{\mathbf{M}}$ then there is a partial interpretation I on a model structure $\mathscr{M} = \langle \mathscr{K}, \mathscr{R}, \lambda \rangle$ with $\mathscr{K}$ finite such that for some $\alpha \in \mathscr{K}'$, $\Gamma = \{A / I_\alpha(A) = T\}$.
Proof. This follows immediately from *L5.5* and *L5.6*.

T5.1. **CS** has the finite model property. That is, if $\{A\}$ is simultaneously satisfiable then $\{A\}$ is simultaneously satisfiable on a model structure $\langle \mathscr{K}, \mathscr{R}, \lambda \rangle$ with $\mathscr{K}$ finite.
Proof. Let $r(A) = j$, and let **M** consist of the sentence parameters that occur in A. By *T6.12* of [12] (which is easily adapted to the sentential case), if $\{A\}$ is simultaneously satisfiable, $\{A\}$ is consistent; hence $\{A\}$ is a consistent subset of $F^j_{\mathbf{M}}$ with **M** finite. By *L5.7* and *T4.2*, it follows that $\{A\}$ is simultaneously satisfiable on a finite model structure.

For more familiar modal logics, decidability is an immediate consequence of the finite model property. This is not so for **CS**, however, since an s-function on a finite model structure is not itself a finite object; f(A,α) must be defined for infinitely many formulas A. The following lemma, however, guarantees that the interpretations on a finite model structure can be coded by means of a finite amount of information.

L5.8. Let v be a valuation of a morphology **M** on a finite model structure $\mathscr{M} = \langle \mathscr{K}, \mathscr{R}, \lambda \rangle$. Corresponding to each s-function f on **M** and $\mathscr{M}$ such that $\langle$v,f$\rangle$ is an interpretation of **M** on $\mathscr{M}$ there is a unique family $\{\leqslant_\alpha\}_{\alpha \in \mathscr{K}}$, of ordering relations such that (1) $\leqslant_\alpha$ is a binary relation on $\{\beta \,/\, \alpha\mathscr{R}\beta \text{ or } \beta = \lambda\}$, (2) $\leqslant_\alpha$ is isomorphic to $\leqslant$ on some initial segment of the nonnegative integers, and (3) α is the least and λ the greatest member of $\{\beta \,/\, \alpha\mathscr{R}\beta \text{ or } \beta = \lambda\}$ under $\leqslant_\alpha$.
Proof. Given I, let $\leqslant_\alpha$ be defined as in *D4.2.* The proof of *L4.5* shows that $\leqslant_\alpha$ meets conditions 1–3. And the proof of *L4.7* shows that where $\{\leqslant_\alpha\}_{\alpha \in \mathscr{K}}$, is thus defined, f($A$,$\alpha$) is the least β such that $\alpha\mathscr{R}\beta$ or $\beta = \lambda$ and $I_\beta(A) = T$. From this it follows that every family $\{\leqslant_\alpha\}_{\alpha \in \mathscr{K}}$ of orderings meeting conditions 1–3, together with a valuation v, determines a unique interpretation.

T5.2. **CS** is decidable.
Proof. Let **M** be a finite morphology. It is clear that if Γ is simultaneously satisfiable in a model structure $\mathscr{M}$, and $\mathscr{M}$ and $\mathscr{M}'$ have the same cardinality, then Γ is simultaneously satisfiable in $\mathscr{M}'$. So let $\mathscr{K}^i = \{j \,/\, 1 \leqslant j \leqslant i \text{ or } j = \lambda\}$, where λ is any fixed object that is not a nonnegative integer. (We assume von Neumann's definition of the nonnegative integers, so that j is a set containing exactly j members.) In view of *T5.1* and the semantic completeness of **CS**, to recursively enumerate the formulas of **M** that are nontheorems of **CS** it will suffice to recursively enumerate the interpretations of **M** on model structures of the form $\langle \mathscr{K}^i, \mathscr{R}, \lambda \rangle$. But for each i the number of binary relations $\mathscr{R}$ on $\mathscr{K}^i$ is finite, and for each reflexive relation $\mathscr{R}$ on $\mathscr{K}^i$ the number of valuations v on $\langle \mathscr{K}^i, \mathscr{R}, \lambda \rangle$ is finite. The desired recursive enumeration of interpretations on finite model structures can then be produced by first listing the interpretations of **M** on model structures in the set $\{\langle \mathscr{K}^1, \mathscr{R}, \lambda \rangle \,/\, \mathscr{R} \text{ is a binary reflexive relation on } \mathscr{K}^1\}$, then those in the set $\{\langle \mathscr{K}^2, \mathscr{R}, \lambda \rangle \,/\, \mathscr{R} \text{ is a binary reflexive relation on } \mathscr{K}^2\}$, and so forth.

In conclusion, we may note that the proof of decidability did not rely essentially on specific assumptions about the alethic modality associated with the conditional connective—or, what comes to the same thing, about structural conditions that may be placed upon the relation $\mathscr{R}$. As in [12], we may suppose that this modality is that of von Wright's **M**, but our argument would have worked equally well for other modalities such as **S4** and **S5**. Our method of proving decidability will therefore apply to these systems of modal logic as well as to **CS** and **M**. In fact, when used directly on such systems this argument provides a uniform way of showing that the systems of logic which possess Henkin-style completeness proofs with respect to Kripke-like semantic interpretations all have the finite model property. Since in most cases only a small amount of detail needs to be added to the completeness proof to secure the stronger result, this method of proving decidability may be worth adding to the general methods that already have been described in the literature.

References

[1] Anderson, A. and N. Belnap, Jr. 1962. The pure calculus of entailment. *The journal of symbolic logic* 27: 19–52.
[2] Åqvist, L. 1971. *Modal logic with subjunctive conditionals and dispositional predicates.* Uppsala: Filosofiska Föreningen och Filosofiska Institutionen vid Uppsala Universitet.
[3] Fitch, F. 1952. *Symbolic logic.* New York: Ronald Press.
[4] Gabbay, D. 1970. Selective filtration in modal logics I. *Theoria* 36: 323–30.
[5] Lewis, D. 1973. *Counterfactuals.* Cambridge: Harvard University Press.
[6] Manor, R. 1971. A semantic analysis of conditional assertion and necessity. Unpublished manuscript, Virginia Polytechnic Institute and State University, Blacksburg, Virginia.
[7] Montague, R. 1968. Pragmatics. In *Contemporary philosophy,* R. Klibansky, ed., pp. 102–21. Florence: La Nuova Italia Editrice. Reprinted in R. Thomason, ed., *Formal Philosophy: Selected Papers of Richard Montague* (New Haven: Yale University Press, 1974).
[8] Scott, D. 1970. Advice on modal logic. In *Philosophical problems in logic: some recent developments,* K. Lambert, ed., pp. 143–73. Dordrecht: D. Reidel.
[9] Segerberg, K. 1968. Decidability of four modal logics. *Theoria* 34: 21–25.
[10] Shukla, A. 1970. Decision procedures for Lewis system S1 and related modal systems. *Notre Dame journal of formal logic* 11: 141–80.
[11] Stalnaker, R. 1968. A theory of conditionals. In *Studies in logical theory,* N. Rescher, ed., pp. 98–112. Oxford: Basil Blackwell.

[12] Stalnaker, R., and Thomason, R. 1970. A semantic analysis of conditional logic. *Theoria* 36: 23–42.

[13] Thomason, R. 1970. A Fitch-style formulation of conditional logic. *Logique et analyse* 13: 397–412.

[14] ——. 1973. Philosophy and formal semantics. In *Truth, syntax, and modality,* H. Leblanc, ed., pp. 294–307. Amsterdam: North-Holland Publishing Co.

12: JOHN MYHILL

Levels of Implication[1]

Fitch's program in mathematical logic may be summarized as follows: Self-reference (and unstratified discourse generally) is not only legitimate but necessary for philosophical purposes. Hence the theory of types is unacceptable as a definitive codification of mathematical reasoning. The problem is to find a demonstrably consistent formalism which will include the full comprehension schema. This necessitates an abandonment of the law of excluded middle, or, to put it in another perhaps more suggestive way, if we define "proposition" as anything which is either true or false, then we must give up the requirement that there be an effective test to find whether a given symbolic expression does or does not express a proposition.

This program was carried out in a series of papers in the *Journal of Symbolic Logic,* of which the most important from our point of view are [1], [2], and [4]. In [1] the "positive" portion of the logic is developed, i.e., the theory of (intensional) identity, the combinators, conjunction, disjunction, existence, and the ancestral. The *axioms* are all expressions $[a = a]$, and the obvious introduction *rules* for each of the notions mentioned are included. In particular, abstracts are defined in terms of combinators in such a way that $[a \epsilon \hat{x}\phi x]$ is a theorem if ϕa is. Since the corresponding elimination rule is a derived rule, we have the effect of a full comprehension schema. Consistency (in Post's sense, since we have no negation) is trivial, since if a and b are different expressions there is no axiom or rule that would give $[a = b]$ as a theorem.

In [2] the system is extended to include negation and universal quantification. The axioms are now $[a = a]$ for each expression a and $\sim[a = b]$ for any two distinct expressions a and b; and for each of the notions mentioned we postulate the obvious *negative introduction rules.* For example, $\sim[a \,\&\, b]$ is a consequence of $\sim a$, and is also a consequence of $\sim b$ (thus, in view of the rule that $\sim\sim a$ is a consequence of a, we can dispense with disjunction as a primitive notion and define $[a \vee b]$ as short for $\sim[\sim a \mathbin{\&} \sim b]$). Again, if the above mentioned definition of abstracts in terms of combinators is adopted, the (derived) rule of negative abstraction introduction reads: $\sim[a \in \hat{x}\phi x]$ is a consequence of $\sim\phi a$. Universal quantification $(x)\,\phi x$ is defined as short for $\sim(\exists x)\sim\phi x$ (which is in turn short for $\sim E(\hat{x}\sim\phi x)$, where E (existence) is a primitive idea); and the (derived) rule of universal quantifier introduction reads: $(x)\phi x$ is a consequence of the denumerably many premisses $\phi\, a_1$, $\phi\, a_2, \phi\, a_3, \ldots$ where $a_1, a_2, a_3 \ldots$ are all the expressions there are. The primitive rule from which this is derived is negative existence introduction: from $\sim(ba_1), \sim(ba_2), \sim(ba_3), \ldots$ infer $\sim(Eb)$. This rule is a nonfinitary one, and hence the system is not a "formal" one in the usual sense of the word: its set of theorems is a Π^1_1 set, not a recursively enumerable set. The proof of consistency is, however, an almost trivial matter: one regards proofs as (in general infinite) well-ordered sequences of expressions, and shows by straightforward induction that if a contradiction occurs in a proof, say a in one line and $\sim a$ in another, then another contradiction must occur earlier. For example, if $a (= Eb)$ is an existential quantification, then it must be obtained by existence introduction (since only introduction rules are used); that is, there must be an expression c such that bc appears earlier in the proof. $\sim a$, i.e., $\sim(Eb)$ must for the same reason be obtained by negative existence introduction, i.e., for each expression d, the expression $\sim(bd)$ must appear earlier in the proof. In particular, the expression $\sim(bc)$ must so appear; so there is a shorter contradictory proof than the one originally given.[2] The proof of closure under the obvious elimination rules (e.g., if $[a \mathbin{\&} b]$ is a theorem, so are a and b; if $(x)\phi x$ is a theorem, so is ϕa for each expression a, etc.) is no harder.

Let us see what happens on this basis to e.g. the Russell paradox. Let R be short for $\hat{x} \sim [x \in x]$; then (by abstraction introduction and abstraction elimination) $[R \in R]$ is a theorem if and only if $\sim[R \in R]$ is. From this, by the consistency result, neither of these two expressions is a theorem, that is, $[R \in R]$ is a proposition which is neither true nor false, or alternatively, if the suggestion made at the beginning of this paper is adopted, it is not a proposition at all. This is in accordance with what common sense would say (if we accept the doubtful premiss that "common sense" admits the unrestricted comprehension schema), and it was obtained with a minimum of effort and very gracefully.

Matters are otherwise, however, if we admit *implication* into the system. Implication is necessary if we are to develop a theory of real numbers; we are at every trick and turn having to make statements like "f is a (real-valued) function of a real variable," i.e., "for every x, *if* x is a real number, so is $f(x)$." The example $f(x) \equiv x$ shows that the representation of $[p \supset q]$ by $[(\sim p) \vee q]$ won't work; for "f (the identity function) is a function of a real variable" then comes out to "(x) ($\sim$(x is a real number) or x is a real number)," i.e., the set of real numbers is a decidable (or in Fitch's terminology a *definite*) set. A straightforward application of the Richard paradox shows that this is not so; for if a set is decidable, we can enumerate its members by a function definable within the system and then diagonalize to obtain a contradiction if (as in this case) the diagonal object itself belongs to the given set. For this reason, in [3] and other papers devoted to the development of elementary analysis in the system of [2], Fitch takes such notions as "f is a function of a real variable" as a notion of the *metalanguage*; f is said to be such a function if, roughly speaking, whenever "a is a real number" is a theorem, so is "$f(a)$ is a real number."

Such a procedure is to say the least, inelegant; instead of embedding part of classical analysis in the theory constructed, we have embedded it in its metalanguage; and it was with a view to mending this situation, amongst other things, that Fitch extended his system further in [4] to include a primitive notion of implication. The result is a nonfinitary natural deduction system; it was later finitized in the book [5], which we shall take henceforth as the definitive formalization of Fitch's logic.[3] The reader interested in following the details of what follows should have Fitch's book at hand.

The problem to be solved in [4] and [5] is as follows: The two essential properties of implication from which, in most natural deduction formalizations, the others are deduced are (1) the principle of *modus ponens,* that b is a consequence of a and $[a \supset b]$ and (2) the *deduction theorem* (or *rule of conditional proof*), that if b can be deduced from a, then $[a \supset b]$ is true. However, the infamous *Curry paradox* shows that *we cannot combine these two principles with the full comprehension schema,* even in systems without negation—hence *the logical paradoxes cannot be avoided by merely dropping the principle of excluded middle.* At the risk of boring the reader, I recapitulate the argument of the paradox. Let $*$ be an arbitrary proposition, perhaps an "absurd" proposition like $[1 = 0]$, and define $K = \hat{x}\ [[x \in x] \supset *]$. Assume $[K \in K]$; then by abstraction elimination $[[K \in K] \supset *]$, and by *modus ponens,* $*$. The assumption $[K \in K]$ has thus led to the conclusion $*$; so, by the deduction theorem,

(1) $[K \in K] \supset *$,

independently of any assumptions. By abstraction introduction,

(2) $[K \in K]$,

and by *modus ponens* a second time, $*$ is a theorem. Thus whatever we do, we can't combine the two abstraction principles and the two implication principles in the same system, at the cost of inconsistency. Fitch's procedure, faced with this dilemma (of having to abandon one of four principles, each of which is apparently completely certain) is: he keeps all four, but he then hedges them about with an ad hoc restriction on proofs, which amounts, roughly speaking, to refusing to allow a subordinate proof, having p as a hypothesis, to appear as an item in the proof of any proposition p. There are various forms of this restriction: one is given in [**4**], and two others in the book [**5**]. They have in common that consistency is preserved (by a much more involved form of the consistency proof outlined above) but that some very evident closure properties are lost in exchange—for example, the resulting systems are not closed under modus ponens nor, in some cases, under conjunction introduction.

While one form of the restriction [**5**, p. 107] is *proof-theoretically* natural, in the sense that it looks like just what one would need to keep contradictions at bay, no form of it known to us is *philosophically* natural, in the sense of being based on a clear analysis of what implication *means.* They appear as *merely formal devices* to avoid inconsistency. The central point of this paper is an attempt to escape the Curry paradox in a more natural and philosophically defensible way.

The paradox certainly establishes that we can't at the same time have full comprehension and have one notion of implication which does what we expect of it. We shall suggest that the resolution lies in admitting many distinct notions of implication, in fact a whole infinite hierarchy of them. In fact once we have got away from explicating "$[p \supset q]$" as "$[(\sim p) \vee q]$," which simply doesn't give us the kind of implication we need in mathematics, what do we mean by saying that a proposition p implies a proposition q? Presumably that some kind of inference leads from p to q. What kind of inference? We can't say valid inference without circularity for the notion of validity itself presupposes implication. (The relation R between premises and conclusion is *valid* if p and pRq *implies* q.) One suggestion that might be offered is that the inference in question be conducted in accordance with certain *rules.* Our point is that a vicious circle results, and is shown by the Curry prardox to result if amongst these rules are the rules (specifically the rules of modus ponens and conditional proof) governing the very sort of implication we are trying to explain. Let us consider first the rules R consisting of all those stated in [5],[4] except the implication rules $R6$ and $R7$. "q is obtainable from p by the rules R" is a perfectly clear relation between two propositions; in fact is is *explicitly definable* in the system of [5]. Call it $\supset_1$; then the system consisting of [5] without $R6$ and $R7$ is closed under modus ponens in the sense that if p and $[p \supset_1 q]$ are theorems, so is q; and under the deduction theorem, in the sense that if there is a deduction of q from p, there is a proof of $[p \supset_1 q]$. However, if we now add the rules of modus ponens and conditional proof for $\supset_1$, nothing prevents the deduction of the Curry paradox. But this is scarcely surprising! – the rule of conditional proof is not valid on the given interpretation! If there is a deduction of q from p *using modus ponens and conditional proof for* $\supset_1$, this gives us no right at all to infer $[p \supset_1 q]$, which says there is a deduction of q from p *not* using these rules. However if there *is* such a deduction, then *another* relation between p and q obtains, likewise explicitly definable in the original system – call it $[q \supset_2 q]$. In general, $[p \supset_1 q]$ means that q is deducible from p without using any implication rules, and $[p \supset_{n+1} q]$ means q is deducible from p using only the rules for $\supset_1, \supset_2, \ldots, \supset_n$. We conclude by giving a rigorous formalization of a system which incorporates these ideas, i.e., we suppose that the reader has a copy of [5], in front of him, and we tell him what modifications he should make.

The *expressions* of the system are the same as the expressions of the system S of [**5**], except that instead of the symbol $\supset$ we have now an infinite number of symbols $\supset_1, \supset_2, \supset_3, \ldots$. The *axioms* are the axioms $R1$ and $R2$ of [**5**, pp. 115–16]. The *rules* are the rules of $R3$–$R45$ of [**5**, pp. 116–17, 192, 211], except that the restrictions (both simple and special, pp. 107, 109) are dropped, trivial modifications of $R44$ and $R45$ are made as described in footnote 4, and the implication rules $R6$ and $R7$ (p. 116) are replaced by

> $R6'_n$. $[p \supset_n q]$ is a d.c. (direct consequence) of a subproof **P** having p as its hypothesis and q as its conclusion, provided that in that subproof no use has been made of any of the rules $R6'_m$, $R6''_m$, $R7'_m$ or $R7''_m$ for $m \geqslant n$, and that none of these rules has been used in the proof of any proposition which has been reiterated into **P** or into any proof subordinate to **P**.
> $R6''_n$.. $\sim[p \supset_n q]$ is a d.c. of p and $\sim q$.
> $R7'_n$. q is a d.c. of p and $[p \supset_n q]$.
> $R7''_n$. p and $\sim q$ are each d.c.'s of $\sim[p \supset_n q]$.

In order to make these rules more rigorous, we can attach *indices* to every line of a proof **P** and to each of its subordinate proofs, etc., as follows. The index of an axiom or a hypothesis is 0. The index of a subproof is the maximum of the indices of its items. The index of a line obtained by reiteration is the same as the index of the line from which it was reiterated. The index of a line obtained by one of $R6'_n$, $R6''_n$, $R7'_n$, $R7''_n$ is the maximum of n and the index (indices) of its premises. The index of a line obtained by any other rule is equal to the index of its premise (or the maximum of the indices of its premises, if it has more than one). With this terminology, $R6'_n$ reads as follows: $[p \supset_n q]$ is a consequence of a subproof having p as its hypothesis and q as its conclusion, provided that subproof has index $< n$.

On this basis we can develop logic almost exactly as in [**5**], and mathematics almost exactly as in, e.g., [3]. (For the latter purpose it turns out to be convenient to define "For all real numbers x" to mean "For all x, x is a real number $\supset_1$," and "For all real functions f" to mean "For all f, f is a real function $\supset_2$.")

As to the consistency of this system, I have encountered some surprisingly messy technical obstacles to following the obvious strategy (of proving by induction the consistency of a sequence of systems $\Sigma_1, \Sigma_2, \ldots$ of which the given system, or some non-finitary extension of it, is the union). I feel strongly that this is the result of my own limitations rather than of any incoherence in the underlying notion of a "hierarchy of implications." So I herewith bequeath the task to someone more pertinacious and ingenious than myself. The Curry paradox at any rate dies at the first blow, as follows: Define $K = \hat{x}[[x \epsilon x] \supset *]$; then we start off

$[K \epsilon K]$	hyp,	0
$[[K \epsilon K] \supset_1 *]$	$R21$,	0
$*$	$R7'_1$,	1
$[[K \epsilon K] \supset_2 *]$	$R6'_2$,	2

(the numbers being the indices of the lines); and we have then nowhere to go.

References

[1] Fitch, F. B. 1942. A basic logic. *Journal of symbolic logic* 7: 105–14.
[2] ——. 1948. An extension of basic logic. *Journal of symbolic logic* 13: 95–106.
[3] ——. 1949. The Heine-Borel theorem in extended basic logic. *Journal of symbolic logic* 14: 9–15.
[4] ——. 1950. A further consistent extension of basic logic. *Journal of symbolic logic* 14: 209–18.
[5] ——. 1952. *Symbolic logic.* New York: Ronald Press.
[6] ——. 1957. Quasi-constructive foundations for mathematics. In *Constructivity in mathematics*, pp. 26–36. Amsterdam: North Holland.
[7] ——. 1958. An extensional variety of extended basic logic. *Journal of symbolic logic* 23: 13–21.
[8] Troelstra, A. S. 1971. Notions of realizability for intuistionistic arithmetic, and intuitionistic arithmetic in all finite types. In *Proceedings of the Second Scandinavian Logic Symposium*, pp. 369–404. Amsterdam: North Holland.

PART IV DEONTIC, EPISTEMIC, AND EROTETIC LOGIC

13: JOHN BACON

Belief as Relative Knowledge

The one obvious logical characteristic of belief, as opposed to mere belief, is that it follows from knowledge. Every knowing is a believing: knowledge is belief of some special kind. Thus it was natural to try to spell out just what kind of belief knowledge is. But the task of defining knowledge in terms of belief has proved unexpectedly arduous (Gettier [13] *et seq.*, e.g. Griffiths [14], Roth-Galis [32] and others). And from the point of view of modal logic, belief is the more problematic notion anyhow. The syntactic and semantic characteristics of some simple logics of knowledge are well understood. An obvious move toward clarifying these concepts is therefore to turn the Plato-Gettier problem around. Can belief be defined in terms of knowledge? As a first step in that direction, I will reduce doxastic logic to epistemic logic.

The point of departure for the reduction is the observation that doxastic is to epistemic logic as deontic is to alethic modal logic. Thus Anderson's [1] and Kanger's [18] reductions of deontic to alethic modal logic are transferrable *mutatis mutandis* to epistemic logic.[1] Belief thereby becomes a special kind of knowledge, "relative" knowledge, as we might say in imitation of Smiley's term *relative necessity* [34]:

(d1) $\mathrm{B}p =_{\mathrm{df}} \mathrm{K}(\nabla \vee p)$.

B here stands for *x believes that,* K for *x knows that,* and *p* for any sentence (of various formal systems to be discussed anon). ∇ is a sentential constant which may be given the preliminary heuristic reading *x is mistaken.* Thus an arbitrary knower *x* believes that *p* if and only if *x* knows that, unless he is mistaken, *p.*

If *x* is the first person *I,* then it is natural to suppress the *I know that* in view of the conversational implicature, pointed out by Unger [35], that I often represent myself as knowing (not just as believing) what I assert in a normal stating context. Thus I may represent myself as believing, but not necessarily knowing, that *p* by saying *Unless I'm mistaken, p.* Further questions about the interpretation of ∇ are deferred to §4.

1. Definitions

A sentential logic X is a *normal epistemic logic* if its connectives include, besides all the usual truth-functional ones, two one-place connectives K and M ("maybe, for all *x* knows,"), and if X meets these conditions ($\vdash$ for *provable in* X):

(k0) X includes the entire classical propositional calculus (PC) and is closed under *modus ponens* and substitution for sentence letters.

(k1) $\vdash \mathrm{M}p \equiv \sim\mathrm{K}\sim p$

(k2) $\vdash \mathrm{K}p \supset p$

(k3) $\vdash \mathrm{K}p \;\&\; \mathrm{K}(\sim p \vee q) \supset \mathrm{K}q$

(k4) if $p \supset q$ is a tautology, then $\vdash \mathrm{K}p \supset \mathrm{K}q$.

(k5) $\vdash p \supset \mathrm{K}p$

(k6) $\vdash \sim\mathrm{K}p$

(k7) there is no p such that $\vdash \mathrm{K}p$.

In view of (k7), X cannot be a normal alethic modal logic, and in view of (k2) it cannot be a normal deontic logic, in the senses of Anderson [2] (with connectives identified in the obvious way). (k7) reflects the fact that knowing is contingent, i.e., it is not logically true that anybody knows anything. (k7) entails the rejection of "L-omniscience" (Chisholm [6, p. 200]), the epistemic counterpart of a rule of necessitation. This rejection parallels von Wright's "principle of deontic contingency" [36, p. 11]. (k7) implies (k5). (k6) rejects the demonstrability of skepticism.[2] (k4) permits the interchange of tautological equivalents in modal contexts. In the presence of (k4), (k3) is deductively equivalent to $\mathrm{K}p \;\&\; \mathrm{K}q \supset \mathrm{K}(p \;\&\; q)$. An example of a minimal normal epistemic logic is Lemmon's E1, consisting of (k0)–(k4) [with (k1) in definitional form] [22, p. 183].

Not all systems of epistemic logic hitherto proposed are normal. Let us call a system fulfilling all the above conditions but with (k7) negated an *idealistic*[3] *epistemic logic.* Idealistic epistemic logics turn out to be the same thing as normal alethic modal logics in Anderson's sense (cf. Theorem 4 below). Indeed, M and S4 have been considered by von Wright [37, pp. 29–32, 57] and propounded by Hintikka [17] as systems of epistemic logic.[4]

A system satisfying (k0)–(k3), (k5)–(k7), the denial of (k4) and

(k8) $\vdash Kp \supset K(p \vee q), \vdash Kp \supset K(q \vee p)$

will be called a *serious* epistemic logic. The trouble with (k4) is the existence, familiar even to a solipsist, of at least one person who does not know all the tautological consequences of his knowledge. (Notice that according to normal epistemic logic, whoever knows anything at all has contingent knowledge of all tautologies; cf. Lemmon [22, p. 185].) I am not familiar with any serious epistemic logics in the literature. The absence of (k4) makes them axiomatically messy. Unlike normal and idealistic epistemic logics, serious ones are not Lewis modal systems in Lemmon's sense [22].

A *normal doxastic logic* is a sentential logic with additional connectives B and C (roughly but mnemonically, "x conjectures" or "considers it possible that") such that

(b0) it includes PC and is closed under *modus ponens* and substitution.

(b1) $\vdash Cp \equiv \sim B \sim p$

(b2) $\vdash Bp \supset p$

(b3) $\vdash Bp \ \& \ B(\sim p \vee q) \supset Bq$

(b4) if $p \supset q$ is a tautology, then $\vdash Bp \supset Bq$.

(b5) $\vdash p \supset Bp$

(b6) for no p is $\vdash \sim Bp$.

(b7) for no p is $\vdash Bp$.

(b8) $\vdash Bp \supset Cp$[5].

Moreover, if the doxastic system is to be an extension of epistemic logic,

(kb) $\vdash Kp \supset Bp$.

In view of (k2), the converse of (kb) is ruled out by (b2). An example of a minimal normal doxastic logic is the system C1 attributed by Routley [33, pp. 239f.] to Lemmon.

An *idealistic doxastic logic* is like a normal one except that (b6)–(b8) are negated. Finally, a *serious doxastic logic* is like a normal one except that (b3) is dropped and (b4) negated. While idealistic doxastic logics are also normal deontic logics and Lewis modal systems, normal and serious doxastic logics are neither [in view of (b8)]. Serious doxastic logic is so weak as to be devoid of architectonic interest. Nevertheless, it is my conviction that no looser logic can capture the everyday sense of *believes*. Serious doxastic logic becomes interesting only when supplemented by deontic modalities or by abstraction permitting the formation of *de re* contexts.

2. *Some General Syntactic Theorems*

We are now ready to consider the results ∇X of adding a sentential constant ∇ together with definitions (d1) and

(d2) $\quad \mathrm{C}p =_{\mathrm{df}} {\sim}\mathrm{B}{\sim}p$

to various epistemic logics X. In all these theorems, (b0) and (b1) follow as a matter of course from our definitions. We shall have need of the three following theses:

(k3′) $\quad \vdash \mathrm{K}.p \supset q. \supset .\mathrm{K}p \supset \mathrm{K}q$[6]

in normal and idealistic epistemic logics, by exportation on (k3), and (k4) on the tautology $p \supset q \supset \bar{p} \vee q$.

(t1) $\quad \vdash \mathrm{K}p \,\&\, \mathrm{K}q \supset \mathrm{K}.p \,\&\, q$

in normal and idealistic logics. Distribute K through the tautology $p \supset .q \supset p \,\&\, q$ successively by (k4) and (k3′), and then import.

(k8) $\quad \vdash \mathrm{K}q \supset \mathrm{K}.q \vee p$

in all our systems. For serious logics, it is a defining condition. In normal and idealistic logics, distribute K through the tautology $q \supset q \vee p$ by (k4). Mainly we shall use the special instance $\mathrm{K}\,\nabla \supset \mathrm{B}p$ [by (d1)].

THEOREM 1. If X is a normal epistemic logic, ∇X is a normal doxastic logic.

Proof. Conditions (b0)–(b1) are met as already noted.

(b2) If it were $\vdash \mathrm{B}p \supset p$, then in particular $\vdash \mathrm{B}\,\bar{\bar{\nabla}} \supset \bar{\bar{\nabla}}$, or $\vdash \mathrm{K}\,\nabla \supset \bar{\bar{\nabla}}$ by (k8). But since also $\vdash \mathrm{K}\,\nabla \supset \nabla$(k2), we get $\sim\mathrm{K}\,\nabla$. No special axioms having been laid down for ∇, this could hold only if $\vdash \sim\mathrm{K}p$, contrary to (k6).[7]

(b3) Distribute K through the tautology $\nabla \vee p.\ \&\ .\nabla \vee .\bar{p} \vee q. \supset \nabla \vee q$ by (k4), and through the antecedent by (t1).

(b4) Disjoin ∇ with both sides and distribute K through by (k4).

(b5) follows from (b7), which follows from (k7) by universal instantiation.

(b6) Suppose there were some sentence p such that $\vdash \sim\mathrm{B}p$. Then by (k8) $\vdash \sim\mathrm{K}\,\nabla$, which would lead to the contradiction of (k6) as under (b2) above.

(b8) If it were $\vdash \mathrm{B}p \supset \mathrm{C}p$, then since $\vdash \mathrm{K}\,\nabla \supset \mathrm{K}.\nabla \vee p$ (k8), we should get $\vdash \mathrm{K}\,\nabla \supset \sim\mathrm{B}\bar{p}$ by hypothetical syllogism. But since likewise by (k8) $\vdash \mathrm{K}\,\nabla \supset \mathrm{B}\bar{p}$, $\vdash \sim\mathrm{K}\nabla$ would follow. This could hold only if $\vdash \sim\mathrm{K}p$, contrary to (k6). (Cf. Prior [30, p. 141, (14)–(18)].)

(kb) Distribute K through the tautology $p \supset \nabla \vee p$ by (k4).

THEOREM 2. If X is a serious epistemic logic, ∇X is a serious doxastic logic.

Proof. (b0)–(b2), (b5)–(b8) as in the proof of Theorem 1.

($\sim$b4) Suppose (b4) were true of ∇X. As no special postulates distinguish ∇ from other sentence letters, (b4) could hold only if, more generally, $\vdash_{PC} p \supset q$ implied that $\vdash \mathrm{K}.r \vee p. \supset \mathrm{K}.r \vee q$, or in particular, that $\vdash \mathrm{K}.\bar{p} \vee p. \supset \mathrm{K}.\bar{p} \vee q$. But then by (k8) $\vdash \mathrm{K}p \supset \mathrm{K}.\bar{p} \vee q$, or $\vdash \mathrm{K}p \supset .\mathrm{K}p\ \&\ \mathrm{K}.\bar{p} \vee q$. By (k3) this would give us $\vdash \mathrm{K}p \supset \mathrm{K}q$, in conflict with ($\sim$k4).

(kb) is an instance of (k8).

THEOREM 3. If X is an idealistic epistemic logic, then ∇X is not an idealistic doxastic logic.

Proof. Let ∇X be an idealistic doxastic logic. By definition, ($\sim$b8) is a thesis of ∇X. Thus, by Theorem 1(b8), X violates (k6) so it is no idealistic (nor any other defined type of) epistemic logic. Contrapositively, if *X is* an idealistic epistemic logic, then ∇X cannot be an idealistic doxastic logic.

For the purposes of the next two theorems, we replace (d1) by the Andersonian

(d3) $\mathrm{B}p =_{df} \mathrm{K}.\dot{\nabla} \supset \mathrm{K}\,\dot{\nabla} \supset p$[8].

The revised systems will be called ∇X.

THEOREM 4. If X is an idealistic epistemic logic, ∇X is an idealistic doxastic logic.
Proof. Essentially given by Anderson in [2]. For if X is an idealistic epistemic logic, it is also a normal alethic modal logic. (I omit the proof of this. The only tricky part is deriving Anderson's axiom ~M.p & $\bar{p}$ in X.) Thus Anderson's theorem establishes that ∇X is a normal deontic logic. ∇X accordingly fulfills all the requirements for an idealistic doxastic logic with the possible exception of (~b6) and (~b7). However, (~b7) follows from (~k7) by (kb), and (~b6) from (~b7) by (b8) and (b1). ∇X is thus an idealistic doxastic logic.

THEOREM 5. If X is a normal epistemic logic, then ∇X is not a normal doxastic logic.
Proof. Even if X is only a normal epistemic logic, $\vdash_X$ M.$p \supset$ Kp, and hence $\vdash_{\nabla X}$ M. ∇⊃ K ∇ (cf. Anderson [1, pp. 175f.]). As Prior [29, p. 141, (7)–(13)] and Anderson [1, p. 173, 1-13] have in effect shown, such a thesis is interdeducible with B$p \supset$ Cp. This contradicts (b8), so ∇X is not a normal doxastic logic.

3. Discussion

The idealistic thesis (~b8), B$p \supset$ Cp, is of special interest for sorting out modal systems. It is the substitution of the analogue of (~b8) for (k2), K$p \supset p$, that weakens Lemmon's epistemic logics to deontic logics. From the deontic standpoint, (~b8) and therefore definition (d3) are desirable. For doxastic purposes, on the other hand, (~b8) is problematic. It has the effect of a consistency posltulate on belief sets (cf. Smiley [34, p. 129]): no one can simultaneously believe contradictory opposites. On the dubious assumption that beliefs are not just always conscious but always "on our mind," it is perhaps difficult to imagine that x could at once believe both that p and that ~p, unless x had a very idiosyncratic conception of ~. Even so,

x's conception of ~ is not idiosyncratic

is not a logical but at best an empirical truth. We might, to be sure, consider with Purtill [30, pp. 274f.] such a thesis as

x is rational ⊃ (B$p \supset$ Cp),

or the deonto-doxastic commitment

$$O(Bp \supset Cp),$$

or the *de re* (viz. *de negatione*)

$$Bp \supset \sim[\hat{\delta}B\delta p](\sim),$$

i.e.,

> if x believes p, then x doesn't believe *of* negation that it qualifies p

as candidates for logical truth. But in simple doxastic logic, (~b8) appears objectionable. This suggests taking doxastic logics, distinguished by (b8), as a fourth and lowest tier in Lemmon's hierarchy of modal systems [22, p. 186].[9] Lemmon has since partially filled in such a tier with his C-systems.

In [1], Anderson moved from (d1) to (d3) in order to make (~b8) provable without special postulation. But if we want (b8), then it is natural to rest with the simpler (d1). Anderson's reduction yielded, in addition to the deontically desirable $\vdash Op \supset Pp$, the deontically baffling $\vdash \Box p \supset Op$,[10] the counterpart of our (kb). This called for some fast deontic footwork: "We should understand '$O\alpha$,' that is, as asserting that α is *either* obligatory . . . *or* necessary" [1, p. 182]. In this respect, the Anderson-Kanger definition (d1) suits our purposes more naturally, for (kb) is the rock-bottom doxastic principle with which we began, the entailment of belief by knowledge.

It is interesting that under our treatment, if an S4 or S5 reduction principle holds for K in X, it will hold for B in the corresponding ∇X. The converse of the S4 reduction principle, $BBp \supset Bp$, is not provable from the defining conditions of any of our ∇X systems. It would become provable in all but serious logics, however, if (k4) were strengthened (k4!) to apply to any thesis, whether or not tautologous. In idealistic systems, that would even make the stronger $B(Bp \supset p)$ ("L-*Rechthaberei*") provable. [The closest normal systems could come would be $\vdash Bq \supset B(Bp \supset p)$; cf. (pr) in §4 below.] But while $O(Op \supset p)$ may be a valid deontic principle, one of the aims of a liberal education is to keep $B(Bp \supset p)$ not just invalid but false. Smiley has, to be sure, argued the plausibility of the deductively equivalent $B\sim(p \;\&\; B\sim p)$ as explaining the pragmatic paradox involved in x's asserting $p \;\&\; B\sim p$. But that, I suspect, comes of confusing the objectionable *de dicto* $B(Bp \supset p)$ with the more plausible *de "re"* (*de propositione,* as it were)

x believes of everything he believes that it is true.

4. *Interpretation*

In terms of what, then, have we defined belief? Apart from truth-functional connectives, (d1) makes use of knowledge and ∇. What is ∇? I suggested the informal reading *x is mistaken*, i.e.,

x is mistaken in at least one of his beliefs.

For purposes of interpretation, it is more convenient to consider the contradictory of ∇, which I shall write as Δ. In analogy to ∇, Δ might be read *x is right,* i.e.,

x is right in his beliefs.

Another way of approximating Δ is as the conjunction of all the sentences x believes to be true, x's "corpus of beliefs" as Smiley puts it [34, p. 113]. Epitomizing that as *what x believes,* we have the doxastic counterpart of Kanger's reading of his deontic constant Q as *what morality prescribes* [18, p. 53]. The parallel is not complete, for although $\vdash OQ$, $B\Delta$ is provable only in idealistic logics. From a normal or serious standpoint, what x believes, he believes contingently.

These informal interpretations, apart from their imprecision, have the drawback of explaining ∇ or Δ in terms of x's beliefs. If we are to escape the appearance of circularity, we must do better than that. In terms of Kripkean relational model theory, what Δ says is that the world is *self-accessible,* possible relative to itself. The precise conditions on the doxastic accessibility relation characterizing Δ in ∇X will, of course, depend on the choice of the underlying epistemic logic X. Routley has, it is true, provided semantics for the minimal normal epistemic logic E1 [33, p. 254]. From here on, however, I will concentrate on the more thoroughly studied normal epistemic logic E2 (Lemmon [22, p. 182]), viz. (k0)–(k3) [(k1) taken definitionally] together with the strengthened rule

(k4!) if $\vdash p \supset q$, then $\vdash Kp \supset Kq$.

As we shall see in the next section (Theorems 6–7), the doxastic semantics induced for ∇ E2 by the Kripkean semantics for E2 [19] are of the following sort (essentially given by Routley [33, pp. 240f.][11]).

A C2′ *model structure* is a triple $\langle \mathscr{K}, \mathscr{N}, \mathscr{S} \rangle$, where $\mathscr{K}$ is a non-empty set (of "worlds"), $\mathscr{N} \subseteq \mathscr{K}$ (*normal* worlds), and $\mathscr{S}$ (doxastic accessibility) is a two-place relation on part of $\mathscr{K}$ ($\mathscr{S} \subseteq \mathscr{K}^2$) which is reflexive on its range ("right-reflexive"). Let h, i, j, k be variables ranging over $\mathscr{K}$. A *valuation* on a doxastic model structure is a two-place function V assigning each sentence or connective Γ a value $V_h(\Gamma)$ in each world h. In the case of a sentence p, $V_h(p) \in \{T,F\}$. Besides treating truth-functional compounds in the usual way for each world, valuations are also subject to the following truth conditions for belief sentences:

$$V_h(Bp) = T \text{ iff } h \in \mathscr{N} \text{ and } V_k(p) = T \text{ for all } k \text{ such that } h \mathscr{S} k.$$

C is characterized dually. Finally,

$$V_h(\Delta) = T \text{ iff } h \mathscr{S} h.$$

A sentence p is C2′-*valid* iff $V_h(p) = T$ for all h under all valuations V on all C2′ model structures. A set of sentences is C2′-*satisfiable* iff they all get T in some world under some valuation on some C2′ model structure.

A useful auxiliary semantic notion is that of the doxastic alternative set of a given world h: $\mathscr{A}(h) = \{k: h \mathscr{S} k\}$. With the help of $\mathscr{A}(h)$, we may characterize B directly in Montague style [26, p. 118] by

$$V_h(B) = \{J: h \in \mathscr{N} \;\&\; \mathscr{A}(h) \subseteq J \subseteq \mathscr{K}\}.$$

If we identify sets with their characteristic functions and the proposition I(p) expressed by p with $\lambda h V_h(p)$ or $\{h: V_h(p) = T\}$, then $V_h(Bp) = V_h(B)(I(p))$ and

$$I(\Delta) = \{h: h \mathscr{S} h\}.$$

The connection between this formal interpretation and our informal reading of Δ shows itself syntactically in the following theses. That x's beliefs all be true in h is a necessary condition for $h \mathscr{S} h$:

(δ1) $\quad \vdash \Delta \supset (Bp \supset p)$.

x's rightness is also a sufficient condition for $h \mathscr{S} h$, *provided* that h is normal:

(t2) $\vdash \mathrm{B}q \supset [\forall p(\mathrm{B}p \supset p) \supset \Delta]$.

Here Bq, x's having a belief (in h), makes h normal. [(t2) is not, of course, a thesis of ∇ E2 as such, but of a straightforward extension to include propositional quantifiers.] The proof of (t2) involves

(δ2) $\vdash \mathrm{B}q \supset \mathrm{B}\Delta$,

itself an interestingly questionable thesis.

Kripke introduced abnormal worlds into epistemic model structures [20, p. 210] in order ot meet (k7). In the abnormal worlds, K behaves as a *falsum* connective. In C2′ model structures the abnormal worlds have a similar function. Because B is *falsum* there, (b7) is satisfied. But for Kripke, every normal world has an alternative, or is *open,* as I shall say. As Makinson has since shown [24], the model condition of openness corresponds to (~b8) (which holds trivially for abnormal worlds). Lemmon accordingly modeled his sytem C2 [= E2 minus the axiom (k2)] by admitting "closed" (alternativeless) normal worlds to C2 model structures [21, pp. 56f.]. (They are admitted simply by omitting an explicit openness condition on $\mathscr{N}$.[12]) In the closed normal worlds B acts as a *verum* connective, for if $\mathscr{A}(h) = \Lambda$, then $\mathscr{A}(h) \subseteq J$ trivially. It is the closed normal worlds, then, that undercut the validity of $\mathrm{B}p \supset \mathrm{C}p$ (C being, dually, *falsum*), thereby guaranteeing (b8) as well as (b6).

C2 is a normal doxastic logic, and is included in the doxastic part of ∇ E2. C2′ model structures result from C2 ones when the condition of right-reflexivity is imposed on the doxastic accessibility relation $\mathscr{S}$ (or at least on $\mathscr{N} 1 \mathscr{S}$[13]). As Routley has shown [33, p. 246], C2′ is adequately axiomatized thus:

(b0) PC, *modus ponens*

(b3′) $\vdash \mathrm{B}(p \supset q) \supset (\mathrm{B}p \supset \mathrm{B}q)$

(pr) $\vdash \mathrm{B}q \supset \mathrm{B}(\mathrm{B}p \supset p)$

(b4!) if $\vdash p \supset q$, then $\vdash \mathrm{B}p \supset \mathrm{B}q$.

In other words, C2′ is C2 plus the axiom schema (pr). I call (pr) the (conditional) Prior principle, since Prior proposed its apodosis as a principle of deontic logic [28; 27, p. 225].[14] Like C1 and C2, C2′ is seen to be a normal doxastic logic.

Fisk has explored a modal analogue of (existence-presupposition-) free logic [11]. A somewhat different analogy from his sheds light on our work. Standard quantification theory is to free (empty-domain-) exclusive quantification theory as epistemic logic is to deontic logic. And the move to inclusive quantification parallels the move to normal doxastic logic:

Standard quantification	Epistemic logic
$\mid\!\vdash \forall xFx \supset Fa$	$\mid\!\vdash \mathrm{K}p \supset p$
$\mid\!\vdash \forall xFx \supset \exists xFx$	$\mid\!\vdash \mathrm{K}p \supset \mathrm{M}p$
Free exclusive quantification	**Deontic logic**
$\dashv\!\mid \forall xFx \supset Fa$	$\dashv\!\mid \mathrm{O}p \supset p$
$\mid\!\vdash \mathrm{E}a \supset (\forall xFx \supset Fa)$	$\mid\!\vdash Q \supset (\mathrm{O}p \supset p)$[15]
$\mid\!\vdash \forall y(\forall xFx \supset Fy)$	$\mid\!\vdash \mathrm{O}(\mathrm{O}p \supset p)$[16]
$\mid\!\vdash \forall xFx \supset \exists xFx$	$\mid\!\vdash \mathrm{O}p \supset \mathrm{P}p$
$\mid\!\vdash \forall x\mathrm{E}x$	$\mid\!\vdash \mathrm{O}Q$[16]
$\mid\!\vdash \forall x\mathrm{E}x$	$\mid\!\vdash \mathrm{P}Q$
(Universally free) inclusive quantification	**Normal doxastic logic**
$\dashv\!\mid \forall xFx \supset Fa$	$\dashv\!\mid \mathrm{B}p \supset p$
$\dashv\!\mid \forall xFx \supset \exists xFx$[17]	$\dashv\!\mid \mathrm{B}p \supset \mathrm{C}p$[18]
$\dashv\!\mid \exists x\mathrm{E}x$	$\dashv\!\mid \mathrm{C}\Delta$[18]
$\mid\!\vdash \exists xGx \supset (\forall xFx \supset \exists xFx)$	$\mid\!\vdash \mathrm{C}q \supset (\mathrm{B}p \supset \mathrm{C}p)$[19]

The possibility envisioned in free logic that $\mathrm{V}(a) \notin \mathfrak{D}$, that a might fall outside the domain of values of x, parallels the possibility in deontic logic that $h \notin \mathscr{A}(h)$, that h might lie outside its alternative set, i.e., that not $h\ \mathscr{S}h$. And the possibility countenanced in inclusive quantification theory that $\mathfrak{D}$ might be empty corresponds to the possibility in doxastic logic that $\mathscr{A}(h)$ might be empty, i.e., h closed. Self-access for the world at hand thus plays the same role in deontic and doxastic logic that existence plays in free logic. Just as free logic drops the assumption that a exists, so deontic logic drops the assumption that $h\ \mathscr{S}\ h$. And just as universally free quantification ( ) can be defined in terms of existence and standard quantification $\forall$ as $(x)Fx =_{\text{df}} \forall x(\text{E}x \supset Fx)$ (Meyer and Lambert [25, pp. 11ff.], reading E for G), so belief is definable in terms of Δ and knowledge as $\text{B}p =_{\text{df}} \text{K}(\Delta \supset p)$ (d1′).

Quine has remarked that even if we refrain from an integral formulation of inclusive quantification theory, we still have a ready criterion of validity in that theory. Simply weed out all standard quantificational L-truths that get F when $\forall$-constituents are marked T and $\exists$-constituents F throughout [31, p. 177]. In other words, the valid inclusive sentences are the intersection of the standard L-truths and the sentences valid under the interpretation of $\forall$ as *verum.* Similarly, the normal doxastic logic C2 is the intersection of Lemmon's deontic D2 with a system in which B is *verum.*[20] The admission of doxastic model structures with closed normal worlds thus has a similar effect as the admission of quantificational models with empty domains.

5. *Semantic Theorems*

From a semantic standpoint, as we have seen, Δ is more natural to work with than ∇. In fact, I chose ∇ and the disjunctive form of (d1) only to increase the apparent plausibility of (k8) in serious epistemic logics. As we are now working with the normal epistemic logic E2, it will be convenient to replace (d1) by

(d1′) $\quad \text{B}p =_{\text{df}} \text{K}.\Delta \supset p$

[and (k3) by (k3′)]. The result of adding the sentential constant Δ and definitions (d1′) and (d2) to E2 will be called ΔE2. Obviously, ΔE2 is equivalent to ∇ E2.

The object of this section is to determine the doxastic fragment(s) of ΔE2. Generalization to Lemmon's E3 and E4 and Kripke's E5 [19, p. 209] is left to the reader. We want, then, to axiomatize the part of ΔE2 that involves no symbols other than sentence letters, truth-functional connectives, B's, C's {and Δ's}. Call that part the B- {B-Δ-} fragment of ΔE2. As already remarked, the B-fragment of ΔE2 is C2′ (Corollary 7.2 below). We establish this syntactic result indirectly via the semantics of the respective systems.

Kripke gives two different definitions of E2 model structures, both of which give rise to the same notion of validity [19, pp.

210f.]. The simpler and epistemically more natural definition equates normal worlds with those on which the epistemic accessibility relation $\mathscr{R}$ is reflexive. According to the more complex definition, $\mathscr{R}$ is reflexive on $\mathscr{N}$ (normal worlds), but possibly also on some abnormal worlds. What both definitions have in common is that $\mathscr{R}$ is reflexive on $\mathscr{N}$. It doesn't really matter epistemically how $\mathscr{R}$ behaves in $\mathscr{K}-\mathscr{N}$, since K is always a *falsum* connective there anyhow. For our purposes, the most serviceable conception of an E2 model structure is a variant of Kripke's second one in which $\mathscr{R}$ is reflexive on *all* of $\mathscr{K}$.[21] An E2 *model structure* is accordingly a triple $\langle \mathscr{K}, \mathscr{N}, \mathscr{R} \rangle$, where $\mathscr{K}$ is a nonempty set, $\mathscr{N} \subseteq \mathscr{K}$, and $\mathscr{R}$ is a two-place relation which is reflexive on $\mathscr{K}$. Truth conditions for K are exactly analogous to those for B (§4 above, reading $\mathscr{R}$ for $\mathscr{S}$). Satisfiability and validity are also defined similarly. It is clear that this specialization of E2 model structures does not affect Kripke's proof of the adequacy of E2, for a sentence will be satisfied by a valuation on all self-accessible abnormal worlds just in case it is satisfied in all abnormal worlds of any sort.

Now, how can we go about imbedding a C2′ model structure in an E2 one? In view of (kb), the two obvious requirements are that $\mathscr{S} \subseteq \mathscr{R}$ and that epistemically normal worlds all be included among the doxastically normal ones. (d1′), however, forces us to equate doxastic with epistemic normality, for there could be no world where K was *falsum* while $K(\Delta \supset p)$ was true. For semantic purposes nothing is thereby lost.[22] Our task, then, is to define a relation $\mathscr{S}$ as a subrelation of $\mathscr{R}$ in such a way that $\mathscr{S}$ is right-reflexive. That is very easily done by arbitrarily restricting the range of $\mathscr{R}$ to get $\mathscr{S}$, i.e., let $\mathscr{S} = \mathscr{R} \upharpoonright \mathscr{J}$ for an *arbitrary* $\mathscr{J} \subseteq \mathscr{K}$. Then $I(\Delta) = \mathscr{J}$. The lack of conditions upon $\mathscr{J}$ is the semantic counterpart of the lack of axioms for Δ in ΔE2. Compare the arbitrariness of the inner domain $\mathfrak{D}$ of free logic (Church [8, p. 103]; Meyer and Lambert [25, p. 11 n. 10]).

Let *b* and *d* be sentences of C2′ with its stock of atomic sentences enlarged to include Δ. In view of (d1′), these sentences are their own translations into ΔE2.

THEOREM 6. If *b* is E2-satisfiable, then *b* is C2′-satisfiable.
Proof. Say that *b* is satisfied by V_h on the E2 model structure $\langle \mathscr{K}, \mathscr{N}, \mathscr{R} \rangle$. Let $\mathscr{S} = \mathscr{R} \upharpoonright I(\Delta)$ (I having been defined in terms of V). $\langle \mathscr{K}, \mathscr{N}, \mathscr{S} \rangle$ is a C2′ model structure, for $\mathscr{S}$ is right-reflexive, as may

be seen from the following argument. Suppose $i \mathscr{S} k$, i.e., $i(\mathscr{R}\restriction \mathrm{I}(\Delta))k$. Then $k \in \mathrm{I}(\Delta)$; and $k \mathscr{R} k$ by definition of $\mathscr{R}$. Thus $k(\mathscr{R}\restriction \mathrm{I}(\Delta))k$, i.e., $k \mathscr{S} k$.

Now let W be a valuation on $\langle \mathscr{K}, \mathscr{N}, \mathscr{S} \rangle$ which coincides with V as far as sentence letters are concerned. We want to show that $\mathrm{W}_h(b) = \mathrm{T}$, which we prove as a special case of $\mathrm{W}_i(b) = \mathrm{V}_i(b)$ for any $i \in \mathscr{K}$. The proof is by induction on the length of b. The *basis case* for sentence letters is covered by stipulation. For $b = \Delta$, $\mathrm{W}_i(\Delta) = \mathrm{V}_i(\Delta)$ follows from the more general $\{h: h \mathscr{S} h\} = \mathrm{I}(\Delta)$, which is proved as follows. Left to right: If $k \in \{h: h \mathscr{S} h\}$, i.e., $k \mathscr{S} k$, i.e., $k(\mathscr{R}\restriction \mathrm{I}(\Delta))k$, then $k \in \mathrm{I}(\Delta)$. Right to left: $k \mathscr{R} k$ by definition of $\mathscr{R}$, so if $k \in \mathrm{I}(\Delta)$, then $k(\mathscr{R}\restriction \mathrm{I}(\Delta))k$, i.e., $k \in \{h: h \mathscr{S} h\}$.

Inductive hypothesis. $\mathrm{W}_j(d) = \mathrm{V}_j(d)$ for d shorter than b. b can take the following forms not already considered in the basis case:

(i) Truth-functional compound. For these V and W coincide by definition.

(ii) Bd. We must establish that $\mathrm{W}_i(\mathrm{B}d) = \mathrm{V}_i(\mathrm{K}.\Delta \supset d)$. Left to right: Suppose $\mathrm{W}_i(\mathrm{B}d) = \mathrm{T}$. Then $i \in \mathscr{N}$ and $\mathrm{W}_k(d) = \mathrm{T}$ for all k such that $i \mathscr{S} k$, i.e., such that $i \mathscr{R} k$ and $k \in \mathrm{I}(\Delta)$, i.e., such that $i \mathscr{R} k$ and $\mathrm{W}_k(\Delta) = \mathrm{T}$. By the inductive hypothesis, we may exchange W for V: $\mathrm{V}_k(d) = \mathrm{T}$ for all k such that $i \mathscr{R} k$ and $\mathrm{V}_k(\Delta) = \mathrm{T}$. Exporting, we get that $\mathrm{V}_k(d) = \mathrm{T}$ if $\mathrm{V}_k(\Delta) = \mathrm{T}$, for all k such that $i \mathscr{R} k$; i.e., $\mathrm{V}_k(\Delta \supset d) = \mathrm{T}$ for those k, and $\mathrm{V}_i(\mathrm{K}.\Delta \supset d) = \mathrm{T}$. Finally, we readjoin $i \in \mathscr{N}$. Right to left: As each step here was equivalential, the argument also works backwards.

(iii) Cd. Treat as $\sim$B$\sim d$.

Those are all the inductive cases, so the inductive argument is finished. In particular, we now know that $\mathrm{W}_h(b) = \mathrm{V}_h(b) = \mathrm{T}$. b has thus been shown to be satisfiable on the C2′ model structure $\langle \mathscr{K}, \mathscr{N}, \mathscr{S} \rangle$. ∎

THEOREM 7. If b is C2′-satisfiable, then b is E2-satisfiable.
Proof. Say that b is satisfied by valuation W for world h on the C2′ model structure $\langle \mathscr{K}, \mathscr{N}, \mathscr{S} \rangle$. Let $\mathscr{R} = \mathscr{S} \cup \{i,j: i = j\}$. As $\mathscr{R}$ is by definition reflexive on $\mathscr{K}$, $\langle \mathscr{K}, \mathscr{N}, \mathscr{R} \rangle$ is clearly an E2 model structure.

Now let V be a valuation on $\langle \mathscr{K}, \mathscr{N}, \mathscr{R} \rangle$ which agrees with W as to atomic sentences. We wish again to show that $\mathrm{V}_i(b) = \mathrm{W}_i(b)$ for arbitrary i, and hence in particular for h. The *basis* is fully covered by stipulation.

Inductive hypothesis. $V_j(d) = W_j(d)$ for d shorter than b.

Case (i) b a truth-functional compound. V and W coincide by definition.

Case (ii) $b = \mathrm{B}d$. To prove: $V_i(\mathrm{K}.\Delta \supset d) = W_i(\mathrm{B}d)$. Left to right: Suppose that $V_i(\mathrm{K}.\Delta \supset d) = \mathrm{T}$. Then $i \in \mathscr{N}$ and $V_k(\Delta \supset d) = \mathrm{T}$ for all k such that $i \mathscr{R} k$, i.e., such that $i(\mathscr{S} \cup \{i,j: i = j\})k$, i.e., such that $i \mathscr{S} k$ or $i = k$. $V_k(\Delta \supset d) = \mathrm{T}$ breaks down into $V_k(d) = \mathrm{T}$ *if* $V_k(\Delta) = \mathrm{T}$ or, by the inductive hypothesis, $W_k(d) = \mathrm{T}$ *if* $W_k(\Delta) = \mathrm{T}$, i.e., *if* $k \in \mathrm{I}(\Delta)$, i.e., *if* $k \mathscr{S} k$. So we have that

(*) $\quad W_k(d) = \mathrm{T}$ if $k \mathscr{S} k$ for all k such that $i \mathscr{S} k$ or $i = k$.

Thus far we have proceeded by equivalential inferences. Now, $i \mathscr{S} k$ implies both that $i \mathscr{S} k$ *or* $i = k$, and that $k \mathscr{S} k$ (right-reflexivity of $\mathscr{S}$). It thus implies for any k that $W_k(d) = \mathrm{T}$. Hence, $W_i(\mathrm{B}d) = \mathrm{T}$. Right to left: We get back to $W_k(d) = \mathrm{T}$ for all k such that $i \mathscr{S} k$. We then prove (*) as follows. If, on the one horn, $i \mathscr{S} k$, then $W_k(d) = \mathrm{T}$ immediately even under the condition that $k \mathscr{S} k$. On the other horn, $i = k$, together with the further condition $k \mathscr{S} k$, implies $i \mathscr{S} k$ by indiscernibility, so that again $W_k(d) = \mathrm{T}$. From (*) we retrace our steps back to the beginning.

That concludes the induction. Accordingly $V_h(b) = W_h(b) = \mathrm{T}$, so b is E-2 satisfiable.

In view of the duality of satisfiability and validity, Theorems 6 and 7 establish

COROLLARY 7.1 $\quad b$ is E2-valid iff b is C2′-valid.

Since Kripke has proved E2 and therewith ΔE2 sound and complete with respect to a variant of E2-validity [19], and Routley has done the same for C2′ with respect to a variant of C2′-validity [34, pp. 245f.], it follows that

COROLLARY 7.2 For any C2′-sentence b, $\vdash_{\Delta \mathrm{E2}} b$ iff $\vdash_{\mathrm{C2}'} b$; i.e., C2′ is the B-fragment of ΔE2.

We turn now to the system C2′Δ, i.e., C2 with the additional sentential constant Δ and the additional axiom schemata

(δ1) $\quad \vdash \Delta \supset .\mathrm{B}p \supset p,$

(δ2) $\quad \vdash \mathrm{B}p \supset \mathrm{B}\Delta.$

(pr) would be redundant as an axiom of C2′Δ: distribute B through (δ1) by (b4!); then by (δ2), Bq implies the resulting antecedent and hence $B.Bp \supset p$. It would be nice to have an axiomatization of $C2'\Delta$ which separated the role of Δ from that of B, such that none of the Δ-axioms depended on the right-reflexivity of $\mathscr{S}$ [as (δ2) does].[23]

As Theorem 7 was proved for sentences of $C2'\Delta$ as well as $C2'$, Corollary 7.1 extends to $C2'\Delta$. It remains to show that

THEOREM 8. $C2'\Delta$ is sound and complete with respect to $C2'$-validity.

Proof. The validity of (δ1)–(δ2) is easily checked: $C2'\Delta$ is sound.

The completeness proof follows Routley [33, pp. 242f., 245f.] and Makinson [24], who should be consulted for details passed over lightly here. I take for granted the usual properties of *saturated* (maximal consistent) sets of $C2'\Delta$-sentences (Lindenbaum sets), in particular *Lindenbaum's lemma* correlating a saturated extension Γ^+ with each noncontradictory set Γ of $C2'\Delta$-sentences, such that $\Gamma \subseteq \Gamma^+$. In the context of this theorem, h and k shall range over saturated $C2'\Delta$-sentence sets.

Routley's Lemma. If $\sim Bb \in h$ and $Bq \in h$ for some q, then $h'(\sim b) = \{\sim b\} \cup \{q\text{: } Bq \in h\}$ is noncontradictory [33, p. 242, Lemma 3].

Now let Γ be any noncontradictory set of $C2'\Delta$-sentences. We want to show that Γ is $C2'$-satisfiable. We do so by constructing a $C2'$ model structure out of saturated $C2'\Delta$-sentence sets. Let $\mathscr{K}$ be the set of all such saturated sets, and $\mathscr{N}$ the set of saturated sets each containing some sentence of the form Bq. Define $\mathscr{S}$ as

$$h \mathscr{S} k \text{ iff for all } p,\ p \in k \text{ if } Bp \in h, \text{ and } \begin{cases} h \in \mathscr{N} \text{ if } k \in \mathscr{N} \\ \Delta \in k \text{ if } k \notin \mathscr{N}. \end{cases}$$ [24]

That $\mathscr{S}$ is right-reflexive may be seen as follows. suppose $h \mathscr{S} k$. Then two cases are possible: (i) $k \in \mathscr{N}$, and (ii) $k \notin \mathscr{N}$.

(i) $k \in \mathscr{N}$. Thus, by definition of $h \mathscr{S} k$, $h \in \mathscr{N}$, i.e., some $Bq \in h$. As saturated sets contain all theses and are closed under *modus ponens*, (pr) and therefore $B(Bp \supset p) \in h$ for all p. Hence, by the definition of $h \mathscr{S} k$, $Bp \supset p \in k$ for all p. k being saturated, this entails that for all p, $p \in k$ if $Bp \in k$. $k \in \mathscr{N}$ if $k \in \mathscr{N}$, and $\Delta \in k$ if $k \notin \mathscr{N}$ trivially, since $k \in \mathscr{N}$. Thus $k \mathscr{S} k$.

(ii) $k \notin \mathscr{N}$. I.e., no $Bp \in k$, so $p \in k$ if $Bp \in k$ trivially for all p. $k \in \mathscr{N}$ if $k \in \mathscr{N}$. And $\Delta \in k$ if $k \notin \mathscr{N}$, by definition from $h \mathscr{S} k$. Thus $k \mathscr{S} k$. In either case, $\mathscr{S}$ is right-reflexive.

As $\mathscr{K}$ is clearly nonempty, $\mathscr{N} \subseteq \mathscr{K}$, and $\mathscr{S}$ is right-reflexive, $\langle \mathscr{K}, \mathscr{N}, \mathscr{S} \rangle$ is a C2′ model structure. Let V be a valuation thereon such that for each sentence letter p of C2′Δ, $V_h(p) = T$ iff $p \in h$ for every h. We wish to prove that V_{Γ^+} satisfies Γ^+ and therefore Γ. We prove that $V_{\Gamma^+}(b) = T$ if $b \in \Gamma^+$ as a special instance of $V_h(b) = T$ iff $b \in h$ for all C2′Δ-sentences b and any saturated sentence set h.

The *basis step* of the induction is covered by the definition of V for atomic sentences other than Δ. As for Δ, suppose that $V_h(\Delta) = T$ but $\Delta \notin h$. Then $h \mathscr{S} h$. By definition of $\mathscr{S}$, $\Delta \in h$ if $B\Delta \in h$; so $B\Delta \notin h$. Thence by (δ2) no $Bp \notin h$, so $h \notin \mathscr{N}$. But by the definition of $h \mathscr{S} h$, $h \notin \mathscr{N}$ leads to $\Delta \in h$, which contradicts our supposition. Conversely, suppose $\Delta \in h$. By (δ1), $Bp \supset p \in h$ for all p; so by saturation, $p \in h$ if $Bp \in h$, for all p. The second two clauses in the definition of $h \mathscr{S} h$ are trivially true. Since $h \mathscr{S} h$, $V_h(\Delta) = T$.

As in the preceding theorems, the *inductive step* boils down essentially to the case where b is of the form Bd. Suppose $V_h(Bd) = T$ but $Bd \notin h$. Then by saturation, $\sim Bd \in h$. Since h is normal, some Bq is in h. By Routley's lemma, $h'(\sim d)$ is noncontradictory. Let r be a sentence letter not occurring in $h'(\sim d)$. Clearly, $H = h'(\sim d) \cup \{Br\}$ is likewise noncontradictory, so by Lindenbaum's lemma we have a saturated set H^+ containing all q such that $Bq \in h$. Since also $Br \in H^+$, $H^+ \in \mathscr{N}$; and h was already in $\mathscr{N}$. Thus, $h \mathscr{S} H^+$. Since $V_h(Bd) = T$, it follows that $V_{H^+}(d) = T$, and by the inductive hypothesis that $d \in H^+$. But as $\sim d$ was by construction in H and hence also in H^+, H^+ is contradictory, which is impossible. Conversely, let $Bd \in h$. If $h \mathscr{S} k$ for any k, then that implies that $d \in k$ or, by inductive hypothesis, $V_k(d) = T$. The supposition furthermore implies that $h \in \mathscr{N}$. Thus $V_h(Bd) = T$.

That completes the inductive argument establishing that $V_h(b) = T$ iff $b \in h$, for all b, h. In particular, then, $V_{\Gamma^+}(b) = T$ iff $b \in \Gamma^+$. Since V_{Γ^+} satisfies Γ^+, it satisfies the subset Γ *a fortiori*. Γ is accordingly C2′-satisfiable.

Since C2′Δ-noncontradictority implies C2′-satisfiability, by duality the C2′-validity[25] of a C2′Δ-argument implies its derivability. C2′Δ is thus not only sound but strongly complete.

Theorem 8 and Corollary 7.1 together entail

COROLLARY 8.1. For any C2′Δ-sentence b, $\vdash_{\Delta E2} b$ iff $\vdash_{C2'\Delta} b$; i.e., C2′Δ is the B-Δ-fragment of ΔE2.

6. Misgivings

Have we really defined belief? The definition is as good as the definiens, which contains Δ, which expresses an arbitrary proposition $\mathscr{J}$, a subset of $\mathscr{K}$. On one construction of epistemic model structures, $\mathscr{J}$ gets identified with the set of doxastically (C2′) self-accessible worlds. What is the *intuitive* content of that? Not much.

It is instructive to compare once again the definition of universally free quantification in terms of standard quantification (end of §4 above). What does E ("exists") mean there? Simply an arbitrary subset $\mathfrak{D}$ of the universe of discourse. As I have argued in [4], free logics, whether or not imbedded in standard logic, do not define the content of the notion of existence. They merely provide the logical framework within which that notion can be used without prejudice. Do they, then, "define" universal free quantification? Well, Lambert's systems "implicitly define" it, and Leblanc and Thomason [20], Meyer and Lambert [25] semantically characterize it. What more is left to do? Nothing but all of ontology.

To return to doxastic logic, we are less interested in Δ in its own right than we are in B, which parallels free ( ) rather than E. We have "implicitly defined" B and semantically characterized it (for certain normal doxastic logics). What more is left? Epistemology.

Apart from (b4!), however, the fly in the ointment is (pr). If $\vdash B(Bp \supset p)$, L-*Rechthaberei,* is intolerable in any fellow, (pr) is about as bad. Since having at least one belief is practically a precondition of being human, by (pr) universal *Rechthaberei* would nevertheless be humanly inescapable. Fortunately for us as practical men, but unfortunately for us as doxastic logicians, it is false.

The rub is that (pr) is intimately involved in our reduction of doxastic logic to normal epistemic logic. (d1/d1′) makes $\mathscr{S}$ right-reflexive on normal worlds, and that is precisely the import of (pr). To remove the right-reflexivity of $\mathscr{S}$ would mean dropping the reflexivity of $\mathscr{R}$, i.e., (k2), thereby turning the epistemic logic into a deontic one. At that point, (d1) loses all intuitive interest so far as I can see. (Witness the theses $Op \supset Bp$, "L-optimism," and, in ΔD4, $Bp \supset OBp$.)

(pr) can be escaped by a hair's breadth by weakening (k4!) back to (k4), thereby returning us to the minimal normal epistemic logic ΔE1. I conjecture the doxastic fragment of ΔE1 to be C1. If so, the appeal of (d1) combined with the repugnance of (pr) would militate in favor of ΔE1 over the formally smoother ΔE2.

If we eschew not only (pr) but also (b4) (which has, remember, the unrealistic consequence that you and I believe all tautologies), we are thrown back upon serious doxastic logic. As Theorem 2 showed, (d1) works even for serious logics. However, the proof made essential use of (k3) and (k8), which, it must be admitted, were assumed ad hoc precisely for that purpose. (k3) and (k8) are really no more plausible than a variety of other particular applications of (k4), e.g., $K(p \vee q) \supset K(q \vee p)$, $K(p \,\&\, q) \supset Kp$, etc. It is difficult to discern any justification for drawing the line where I did. In the absence of a criterion, the most natural courses would seem to be either the exclusion even of (k3) and (k8), or the admission of (k4). The former course would destroy the reductive power of (d1); the latter, the seriousness of ∇X.

The conditions for a serious doxastic logic are all negative, apart from (b0), (b1), and (kb). Considerations essentially like those advanced against (~b8) in §3 can in fact be marshaled against any positive doxastic principle. We are driven ultimately to Cresswell's conclusion that there is no autonomous logic of belief, in the sense that all valid belief sentences devoid of modal connectives other than B and □ are substitution instances of nondoxastic logical truths [9, pp. 12f.]. Only our initial principle (kb) is thus left standing, along with (b1). (d1) could be sustained on the same footing as an equivalence relating belief and knowledge, but no longer as a reductive definition.

7. *"Knowledge" Defined*

After the death scene, dramatic form calls for an upturn. We took off from the Plato-Gettier problem. An unexpected dividend of our logical odyssey is a candidate for that much-sought definition of knowledge in terms of belief. As Chisholm puts the problem,

> "*S* knows that *h* is true" . . . is assumed to tell us three different things:
>
> 1. *S* believes that *h* . . .
> 2. *h* is true . . . And finally, . . .
> 3. ——.

Thus we have a blank to fill. What shall we say of 3? [7, pp. 5f.].

The answer is Δ. Then, of course, 2 becomes superfluous in view of (δ1). We may accordingly define

(d4) $\mathrm{K}p =_{\mathrm{df}} \Delta \mathbin{\&} \mathrm{B}p$.

THEOREM 9. If to C2′Δ we add (d4) and

(d5) $\mathrm{M}p =_{\mathrm{df}} {\sim}\mathrm{K}{\sim}p$,

then C2′Δ becomes a normal epistemic logic.

The proof is left to the reader. As it makes no essential use of Prior's principle, Theorem 9 might be expected to hold of a system C2Δ too, pending an axiomatization thereof. A similar theorem is provable using the slightly weaker definition

(d4′) $\mathrm{K}p =_{\mathrm{df}} \forall q(\mathrm{B}q \supset q) \mathbin{\&} \mathrm{B}p$

with an extended doxastic propositional calculus C2† lacking Δ. However, fortunately or not, universal doxastic rectitude, $\forall q(\mathrm{B}q \supset q)$, is contingently false. (d4) and (d4′) thus lead to de facto, if not demonstrable, skepticism. Theorem 9 notwithstanding, therefore, (d4), like Dawson's definition of "obligation" [10], is of purely formal interest.

References

[1] Anderson, Alan Ross. 1956. The formal analysis of normative systems. Reprinted in *The logic of decision and action,* ed. Nicholas Rescher, pp. 147–213. Pittsburgh: University of Pittsburgh Press, 1967.

[2] ——. 1958. A reduction of deontic logic to alethic modal logic. *Mind* 67: 1–4.

[3] Bacon, John. 1973. Kripke's deontic semantics again. *Notre Dame journal of formal logic* 14: 581f.

[4] ——. 1969. Ontological commitment and free logic. *The monist* 53: 310–19.

[5] ——. 1967. Syllogistic without existence. *Notre Dame journal of formal logic* 8: 195–219.

[6] Chisholm, Roderick. 1963. The logic of knowing. In [32], pp. 189–219.

[7] ——. 1966. *Theory of knowledge.* Englewood Cliffs: Prentice-Hall, Inc.

[8] Church, Alonzo. 1965. Review of Karel Lambert, "Existential import revisited." *Journal of symbolic logic* 30: 101f.

[9] Cresswell, M. J. 1972. Intensional logics and logical truth. *Journal of philosophical logic* 1: 2–15.

[10] Dawson, E. E. 1959. A model for deontic logic. *Analysis* 19: 73–8.
[11] Fisk, Milton. 1969. A modal analogue of free logic. In *The logical way of doing things,* ed. Karel Lambert, pp. 147–84. New Haven: Yale University Press.
[12] Fitch, Frederic B. 1952. *Symbolic logic: an introduction.* New York: The Ronald Press.
[13] Gettier, Edmund L. 1963. Is justified true belief knowledge?" In [32] pp. 35–8.
[14] Griffiths, A. Phillips, ed. 1967. *Knowledge and belief.* London: Oxford University Press.
[15] Hilpinen, Risto, ed. 1971. *Deontic logic: introductory and systematic readings.* Dordrecht: D. Reidel Pub. Co.
[16] ——. 1968. Rules of acceptance and inductive logic. *Acta Philosophica Fennica* 22.
[17] Hintikka, K. Jaakko J. 1962. *Knowledge and belief: an introduction to the logic of the two notions.* Ithaca: Cornell University Press.
[18] Kanger, Stig. 1957. New foundations for ethical theory. In [15], pp. 36–58.
[19] Kripke, Saul A. 1965. Semantical analysis of modal logic II: non-normal modal propositional calculi. In *The theory of models,* ed. Addison, Henkin, and Tarski, pp. 206–20. Amsterdam: North-Holland Pub. Co.
[20] Leblanc, Hugues, and Thomason, Richmond H. 1968. Completeness theorems for some presupposition-free logics. *Fundamenta mathematicae* 62: 125–64.
[21] Lemmon, E. J. 1966. Algebraic semantics for modal logics I. *Journal of symbolic logic* 31: 46–65.
[22] ——. 1957. New foundations for Lewis modal systems. ibid. 22: 176–86.
[23] Levi, Isaac. 1967. *Gambling with truth.* New York: Alfred A. Knopf.
[24] Makinson, David C. 1966. On some completeness theorems in modal logic. *Zeitschrift für mathematische Logik und Grundlagen der Mathematik* 12: 379–84.
[25] Meyer, Robert K., and Lambert, Karel. 1968. Universally free logic and standard quantification theory. *Journal of symbolic logic* 33: 8–26.
[26] Montague, Richard. 1968. Pragmatics. In *Contemporary philosophy/La philosophie contemporaine* I: *logic and foundations of mathematics,* ed. Raymond Klibansky, pp. 102–22. Florence: La Nuova Italia Editrice.
[27] Prior, A. N. 1962. *Formal logic.* 2nd ed. Oxford: The Clarendon Press.
[28] ——. 1956. A note on the logic of obligation. *Revue Philosophique de Louvain* 54: 86f.
[29] ——. 1957. *Time and modality.* Oxford: The Clarendon Press.
[30] Purtill, Richard L. 1971. *Logic for philosophers.* New York: Harper and Row.
[31] Quine, W. V. 1954. Quantification and the empty domain. *Journal of symbolic logic* 19: 177ff.

[32] Roth, Michael D., and Galis, Leon, ed. 1970. *Knowing: essays in the analysis of knowledge.* New York: Random House.
[33] Routley, Richard. 1970. Extensions of Makinson's completeness theorems in modal logic. *Zeitschrift für mathematische Logik und Grundlagen der Mathematik* 16: 239–56.
[34] Smiley, Timothy. 1963. Relative necessity. *Journal of symbolic logic* 28: 113–34.
[35] Unger, Peter. 1972. A defense of skepticism. Lecture given at the Summer Institute in the Theory of Knowledge, Council for Philosophical Studies, Amherst College, 5 July 1972. Different from his published article of the same title.
[36] Wright, Georg Henrik von. 1951. Deontic logic. *Mind* 60: 1–15.
[37] ——. 1951. *An essay in modal logic.* Amsterdam: North-Holland Pub. Co.

14: KATHLEEN JOHNSON WU

Believing and Disbelieving*

Most formalizations of the logic of knowledge and belief are developed with the purpose of explaining the way knowledge and belief statements are actually used. Indeed, Jaakko Hintikka considers it a serious drawback that the results of his system in [4] are applicable to such statements "only insofar as our world approximates, or can be made to approximate, an 'epistemically perfect world' " [5, p. 2]. As a solution, Hintikka has suggested restricting his system in such a way that if q is a logical consequence of p and a person knows (believes) that p, it need not follow that he also knows (believes) that q, unless "$p \supset q$"[1] is a surface tautology at the depth of p [6]. Even with this restriction, however, Hintikka admits that his system would still involve some idealization, as it would be applicable only under the false assumption that everyone "fully understands" whatever he knows or believes.

The problem with Hintikka's proposed solution is that, regardless of how weak the required assumption, it is possible that it is not warranted in a particular case. The advantage of a system I have proposed [8 and 9] is that it involves no restrictive assumptions and, as a result, is applicable without a requirement of logical omniscience, full understanding, or the like. When extra assumptions are useful, they may be explicitly made. For instance, once Hintikka works out the details of his restricted system, it should not be difficult to see how to develop an extension of my system which would, like Hintikka's, be applicable only under the assumption that everyone "fully understands" whatever he knows or believes. This would be done by adopting special rules of usage to correspond to the requirement. The procedure would be the same

*This paper was presented at the annual meeting of the Pacific Division of the American Philosophical Association at San Francisco, 29 March 1974. Douglas Walton was the commentator. I would like to thank him as well as Max Hocutt, Norvin Richards, and Caroline Plath for their helpful remarks.

as that followed in [8 and 9] with respect to the addition of rules of ideal usage to correspond, in one case, to the requirement that everyone is deductively omniscient and, in the other, to the requirement that everyone not only is deductively omniscient but also knows that everyone is, knows that everyone knows that everyone is, and so on.

Without the addition of such rules, however, the system is well suited, indeed, in my view better suited, to deal with problems the correct analysis of which depends on making clear particular assumptions involved rather than on appealing to general standards of deductive behavior. Moore's Paradox and the Prediction Paradox, both of which are dealt with in the following, appear to be two such problems.

I

Moore's problem of saying and disbelieving, often referred to as Moore's Paradox, is that of explaining why the sentence

(1) "*p*, but I don't believe that *p*,"

though logically consistent, seems absurd for someone to say. Hintikka's solution is that if a person says (1), the "general presumption that the speaker believes or at least can conceivably believe what he says" is violated [4, p. 67], because the speaker cannot help but somehow feel that

(2) "$B_a(p\&{\sim}B_ap)$"

is indefensible—where *a* is the first-person, singular pronoun "I"—and, for this reason, cannot believe what he says. Such awareness is inevitable, Hintikka insists, given the simplicity of his formal proof that (2) is indefensible. In [6], because of restrictions he intends to place on (C.BB*), Hintikka notes that this proof may no longer stand. In view of this, he recalls an alternative explanation provided in [4, p. 70] based on a still simpler proof that demonstrates independently of (C.BB*) the indefensibility of

(3) "$(p\&{\sim}B_ap)\ \&B_a(p\&{\sim}B_ap)$."

The simplicity of this proof is taken to indicate that a person cannot conceivably believe that (1). The assumption is that a person cannot conceivably believe a statement which he cannot assert together with his belief in it without making an obviously indefensible statement. The question for Hintikka is: How can (2) be defensible (i.e., true in some logically ideal world) and a person's believing (1) yet be impossible?

Hintikka considers his solution to the problem of saying and disbelieving superior on two counts to both G. E. Moore's and Max Black's. First, he regards it as more economical with respect to the assumptions involved. Secondly, he sees it as an aid, which the others are not, to understanding why it is odd for a person to say

(4) "*p*, but I don't know that *p*."[2]

As Hintikka argues, a person who says (4) cannot know what he says, because

(5) "$K_a(p \& \sim K_a p)$"

can be easily proven indefensible.[3] Therefore, a person's saying (4) violates the general presupposition that a person can conceivably know what he is saying is true. To explain why a person's saying (4) seems less awkward than his saying (1), Hintikka writes that

> other things being equal, it is less unnatural to say something one does not (and cannot) know than to say something one obviously does not (and cannot) believe [4, p. 82].

I leave it to the reader to judge, but it seems that, with regard to the two criteria on which Hintikka considers his solution better than both Moore's and Black's, the following explanation is more satisfactory than his.

On most occasions, if a person is heard saying

(6) "*q*, but I do not believe that *p*,"

it is reasonable to assume that he believes that *q*, but does not believe that *p*. This is particularly true where *p* and *q* have a similar syntax, as in "John is coming" and "Mary is coming." Where *p* and *q* are the same sentence, (6) becomes (1). If this is correct so far, the strangeness of a person's saying (1) can then be explained on the grounds that it gives rise to an obviously absurd assumption: that the person who utters (1) both believes and does not believe that *p*. The oddness of a person's saying (4) may be explained in an analogous manner. On most occasions, if a person is heard saying

(7) "*q*, but I do not know that *p*,"

it is reasonable to assume he knows that *q* but does not know that *p*, particularly where *p* and *q* have a similar syntax. Where *p* and *q* are the same sentence, (7) becomes (4). The strangeness of a person's saying (4) can then be explained on the grounds that it gives rise to an obviously absurd assumption: that the person who utters (4) both knows and fails to know that *p*.[4] The reason a person's saying (4) may seem less absurd than a person's saying (1) is that it is possible to assume that the person uttering (4) merely believes that *p* but does not know that *p*. This would be a perfectly consistent assumption to make. However, if on the basis of a person's saying (1) it were assumed that he knows that *p* but does not believe that *p*, the assumption would be just as absurd as the assumption that the person both believes and does not believe that *p*, because knowledge implies belief.

The oddness of a person's saying

(8) "I believe that *p*, but I do not believe that I believe that *p*"

and

(9) "I know that *p*, but I do not know that I know that *p*"

can be explained by showing that (8) and (9) are special cases, respectively, of (1) and (4). Why a person's saying (9) may seem less absurd than a person's saying (8) can also be understood within the same context.

II

The Prediction Paradox involves the following situation: a person in a position to know forecasts that a particular surprise event will occur within a certain set of days. A person whom the event would affect and for whom the event is to constitute a surprise argues that it cannot occur on the last day, because then he would not be surprised. Since it cannot occur on the last day, it must occur on some day previous to the last, but not on the next-to-last day, because then he would not be surprised, nor on the next-to-next-to-last day for the same reason, and so on. His conclusion is that it cannot occur at all. But the event does occur as predicted and takes him by surprise.

A number of scholarly papers have been devoted to the puzzle. Given the argument, how can the prediction that the surprise event will occur within a certain set of days be fulfilled? In [7], W. V. Quine rejects the supposition that an actual antinomy is involved. His view is that the argument is faulty, because it does not take into account all relevant alternatives. For instance, the part of the argument concluding that the event cannot occur on the last day does not allow in its premisses for the possibility that the event will occur on the last day and be a surprise. Others interpret the prediction differently. Some understand "a surprise event" as meaning an event that is not provably implied by the prediction and the fact that the event does not occur on a previous day. The most successful analysis employing this reading is Frederic Fitch's. In [3], he uses Gödel's technique to handle the self-referential aspect of this interpretation of the announcement and concludes that "the 'paradoxical' prediction is paradoxical only in the rather weak sense of being self-contradictory" [3, p. 163].

Arguing that an analysis of the prediction in terms of self-reference fails to explain how what appears to be a flawless argument is brought to naught, Robert Binkley proposes in [1] to use epistemological and pragmatic concepts similar to those Hintikka uses in his analysis of Moore's Paradox. As a result, Binkley concludes that the prediction belongs to the same family as Moore's Paradox. Both are incredible but possibly true propositions. The prediction is incredible in the special sense that no person to whom the announcement is made can believe it if he is an ideal knower. An ideal knower is understood by Binkley to be a person with the following traits: (i) he follows up all the logical consequences of what he believes and avoids all contradiction; (ii) he believes that he is an ideal knower; (iii) he believes that he believes whatever he believes; (iv) if an event does not occur on a certain day, he believes on the days following that on that day the event did not occur; and (v) if he believes something to be the case on a certain day, he believes on that day that on every following day he will still believe that it is the case.

In my view, Binkley's appeal to an ideal knower is unnecessary and does little to explain the puzzle of the so-called paradox. What is needed, it seems, is not only a reasonable interpretation of what the prediction states but also a careful analysis of the argument and its assumptions. In what follows, such an interpretation and analysis is attempted, in order to show how, in spite of the argument, the prediction can be fulfilled. To state the prediction more concisely, we let **a** be the person for whom the event is to constitute a surprise and who, more or less in desperation, argues that it cannot occur on any of the days in the specified set. We understand by the event's being a surprise that, on the day it occurs, **a** does not believe, beforehand, that it occurs on that day. The prediction can then be seen as stating that the event occurs on exactly one day in a certain set of days, and on the day it occurs **a** does not believe, beforehand, that it occurs on that day.

The part of the argument concluding that the event cannot occur on the last day seems to be based on the following assumptions: (a) if the event occurs on the last day, then it does not occur on any other day in the set; (b) if the event does not occur on a previous day in the set, then on the last day, before the event occurs on that day, if it does, **a** believes that the event does not occur on a previous day in the set; (c) on the last day, before the event occurs on that day, if it does, **a** believes that if the event does not occur on any other day in the set, then it occurs on the last; (d) on the last day, before the event occurs on that day, if it does, if **a** believes that the event does not occur on any other day in the set and also believes that if it does not, then it occurs on the last, he believes that it occurs on the last; and (e) if the event occurs on the last day, then on that day **a** does not believe, beforehand, that it occurs on that day. Assumptions (a) and (e) are true if the prediction is true. The seriousness of the event for **a**, especially in the case of his own hanging, makes (b) and (d) natural assumptions. (c) is related to **a**'s believing the prediction. Since the prediction is made by someone in a position to know, the assumption that on the last day **a** believes that it is true is also at this point in the argument a natural assumption.

The part of the argument concluding that the event cannot occur on the next-to-last day seems to be based on the following assumptions: (f) on the next-to-last day, before the event occurs on that day, if it does, **a** believes that the event does not occur on a later day; and (a′)–(e′) if the next-to-last day is not also the first. If the next-to-last day is also the first, then the assumptions are

(f) and (c′)–(e′). (a′)–(c′) and (e′) are like (a)–(c) and (e), respectively, except for containing the word *next-to-last* where the latter contain the word *last.* If the next-to-last day is also the first, then the similarity between (d′) and (d) is the same. However, if the next-to-last day is not also the first, then (d′) reads as follows: On the next-to-last day, before the event occurs on that day, if it does, if **a** believes that the event does not occur on a previous day in the set, and also believes that it does not occur on a later day in the set, and believes further that if it does not occur on a previous and does not occur on a later day, then it occurs on the next-to-last, he believes it occurs on the next-to-last day. (a′)–(e′) are as natural as (a)–(e); (f) is based on the preceding part of the argument, concluding that the event cannot occur on the last day. The remaining parts of the argument seem to be based on assumptions analogous to the case just considered. All that is needed to obtain them is to change the word *next-to-last* in (a′)–(e′) and (f) for the word *next-to-next-to-last* and then by the word *next-to-next-to-next-to-last,* and so on, depending on how many days there are in the set.

Given the assumptions, the conclusions follow easily. Suppose the event occurs on any day in the set, a contradiction can be readily obtained and the assertion that it cannot occur on that day, therefore, justified.[5] From an examination of the argument as a whole, it is not difficult to see why the event can occur as predicted in spite of the argument. As a result of his argument, **a** concludes that the event cannot occur on any day in the set. As long as he remains convinced of this, he insures the key assumptions on which his argument rests will be false, i.e., (c) and every assumption like (c) which states that on a particular day in the set, before the event occurs on that day, if it does, **a** believes that if the event does not occur on any other day in the set, it occurs on that day.[6]

Appendix

To state the argument precisely, let $\mathbf{p}_1, \mathbf{p}_2, \ldots \mathbf{p}_n$ be the propositions, respectively, that the event occurs on the first day, that it occurs on the second day, . . . that it occurs on the last day. Let $\mathbf{B}_2{\sim}\mathbf{p}_1, \ldots \mathbf{B}_n{\sim}\mathbf{p}_{n-1}$ be the propositions, respectively, that **a** believes on the second day, before the event occurs on that day, if it does, that the event does not occur on the first day, . . . that **a** believes on the last day, before the event occurs on that day, if it does, that the event does not occur on the next-to-last day. The assumptions on which the part of the argument rests that concludes the event cannot occur on the last day can now be stated symbolically as follows: (a) $\mathbf{p}_n \supset (\sim\mathbf{p}_1 \,\&\, \ldots \,\&{\sim}\mathbf{p}_{n-1})$; (b) $(\sim\mathbf{p}_1 \,\&\, \ldots \,\&{\sim}\mathbf{p}_{n-1}) \supset \mathbf{B}_n(\sim\mathbf{p}_1 \,\&\, \ldots \,\&{\sim}\mathbf{p}_{n-1})$; (c) $\mathbf{B}_n((\sim\mathbf{p}_1 \,\&\, \ldots \,\&{\sim}\mathbf{p}_{n-1}) \supset \mathbf{p}_n)$; (d) $(\mathbf{B}_n(\sim\mathbf{p}_1 \,\&\, \ldots \,\&{\sim}\mathbf{p}_{n-1}) \,\&\mathbf{B}_n((\sim\mathbf{p}_1 \,\&\, \ldots \,\&{\sim}\mathbf{p}_{n-1}) \supset \mathbf{p}_n)) \supset \mathbf{B}_n\mathbf{p}_n$; and (e) $\mathbf{p}_n \supset \sim\mathbf{B}_n\mathbf{p}_n$.

To simplify matters, we choose $n = 2$. The part of the argument concluding that the event cannot occur on the last day and the part concluding that it cannot occur on the next-to-last or, in this case, the first day, can be reasonably formulated Fitch-style by proofs 1–12 and 13–21, respectively, which follow below.

Items 1–5, which are the hypotheses of the first proof, correspond, respectively, to (a)–(e). Item 6 is the hypothesis of the subordinate proof. Implication elimination is used to derive 7 from 1 and 6, 8 from 2 and 7, 10 from 4 and 9, and 11 from 5 and 6. Conjunction introduction is used to derive 9 from 3 and 8. 12 follows from the subordinate proof by negation introduction. Reiterations into subordinate proofs are left as understood. Items 13–16, which are the hypotheses of the second proof, correspond, respectively, to (f) and (c′)–(e′). Conjunction introduction is used to derive 18 from 13 and 14. Implication elimination is used to derive 19 from 15 and 18, and 20 from 16 and 17. 21 follows from the subordinate proof by negation introduction.

1 | $\mathbf{p}_2 \supset \sim\mathbf{p}_1$
2 | $\sim\mathbf{p}_1 \supset \mathbf{B}_2\sim\mathbf{p}_1$
3 | $\mathbf{B}_2(\sim\mathbf{p}_1 \supset \mathbf{p}_2)$
4 | $(\mathbf{B}_2\sim\mathbf{p}_1 \,\&\mathbf{B}_2(\sim\mathbf{p}_1 \supset \mathbf{p}_2)) \supset \mathbf{B}_2\mathbf{p}_2$
5 | $\mathbf{p}_2 \supset \sim\mathbf{B}_2\mathbf{p}_2$
6 | | $\mathbf{p}_2$
7 | | $\sim\mathbf{p}_1$
8 | | $\mathbf{B}_2\sim\mathbf{p}_1$

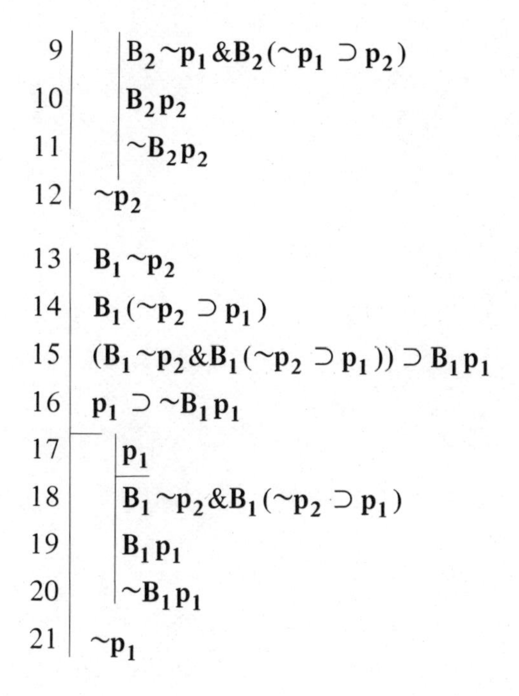

References

[1] Binkley, Robert. 1968. The surprise examination in modal logic. *The journal of philosophy* 65: 127–36.
[2] Fitch, Frederic. 1963. A logical analysis of some value concepts. *The journal of symbolic logic* 28: 135–42.
[3] ——. 1964. A Gödelized formalization of the prediction paradox. *American philosophical quarterly* 1: 161–64.
[4] Hintikka, Jaakko. 1962. *Knowledge and belief.* Ithaca, New York: Cornell University Press.
[5] ——. 1966. 'Knowing oneself' and other problems in epistemic logic. *Theoria* 32: 1–13.
[6] ——. 1970. Knowledge, belief and logical consequence. *Ajatus* 32: 32–47.
[7] Quine, W. V. 1953. On a so-called paradox. *Mind* 62: 65–67. Reprinted in 1966 under the title "On a Supposed Antinomy" in *The ways of paradox*. New York: Random House.
[8] Wu, Kathleen Gibbs Johnson. 1970. A new formalization of the logic of knowledge and belief. (Ph.D. diss., Yale University.)
[9] Wu, Kathleen Johnson. 1973. A new approach to formalization of a logic of knowledge and belief. *Logique et analyse.* 16: 513–25.

15: CARL R. KORDIG

Relativized Deontic Modalities

I

Standard deontic logic is here extended to include relativized deontic modalities. Deontic operators are relativized to sets of one or more persons. Relativized deontic modalities concern what is obligatory or permissible for an individual or group of individuals, as contrasted to what is only "impersonally" obligatory. Such an extension of deontic logic concerns connections between obligations that bind particular persons, on the one hand, and impersonal norms, on the other. In a recent paper Hintikka underscores the need for such a study:

> In this paper we shall disregard the (highly relevant) fact that obligations, and even more so duties, are usually thought of as obliging *some particular person,* and that permissions often are similarly 'personal' . . . the reader will have to think of obligations in completely 'impersonal' terms. . . . Whether, and if so, how, 'personal' obligations and permissions can be defined in terms of the 'impersonal' ones is an interesting question which will not be discussed here [9, p. 60, italics his].

I will discuss this important question here. In current formulations of deontic logic [8, 9] the variables and quantifiers range over individual *acts,* not individual persons. Obligatory acts are "impersonal." There is no way to say either (a) that acts are obligatory for a particular person or group of particular persons or (b) that some acts are obligatory for some, but not all, persons.

Significant philosophical considerations also motivate the extension of deontic logic to include relativized deontic modalities. Distinctions expressed in ordinary language suggest the need for such an extension. The line between what is supererogation and what is not is such a distinction. Saints and heroes have obligations beyond those of ordinary folk. A doctor who truly is a saint has specifically saintly obligations (for an example here, cf. [12, p. 62]. There are analogous distinctions. A head of state has special obligations beyond those of its citizens. A cleric, a judge, a policeman, have special obligations not shared by others.

Logic may be studied by syntactical and/or semantical methods. In section 2, I give relativized deontic modalities a semantics; in Section 3 an axiom system; in Section 4, I apply the logical theory to the philosophical problem of supererogation.

II

In this section I will review, and then extend, a (Hintikka) semantics for deontic modalities to include deontic modalities relativized to persons.

In deontic logic we deal with predicates of individual acts. The letters G, H. . . . are arbitrary predicates. The letters x, y, . . . are arbitrary bound act-individual variables. The letters i, j, k . . . are arbitrary free act-individual variables. The quantifiers are the usual symbols '(∃x)' and '(y)'. Since x, y, . . . are the only bound variables, quantification is only over act individuals. The familiar propositional connectives are '~' (not), '&' (and), 'v' (inclusive or), '⊃' (if-then), and '≡' (if and only if). The operators 'O' (it ought to be the case that), 'P' (it is permissible that), and 'F' (it is forbidden that) express the fundamental deontic modalities. The letters p, q, r, . . . are arbitrary formulae. O and F are defined in terms of P:

(D.1) $Op \equiv \sim P \sim p$;

(D.2) $Fp \equiv O \sim p$.

'p(i/j)' is the formula obtained from p by replacing 'j' everywhere by 'i'. I will use similar notation for bound variables, and for bound and free variables.

The following are rules of formation:

(F.1) An n-adic predicate followed by a pair of parentheses enclosing n free variables is a formula.

(F.2) If p and q are formulae, then so are ~p, (p & q), (p v q), Op, Pp, and Fp.

(F.3) If p is a formula containing the free variable 'i' without the bound variable 'x', then (∃x)p(x/i) and (x)p(x/i) are also formulae.

(F.1)–(F.3) define the formulae of nonrelativized deontic logic. Let us now proceed to provide a semantics for such formulae, using Hintikka's technique of model sets in a model system.

A model set is a set of formulae, a partial description of a logically possible world. The letters μ, μ^*, μ^{**}, etc. stand for model sets. Any set α is a *model set* μ, if and only if:

(C.~) if $p \epsilon \alpha$, then not $\sim p \epsilon \alpha$;

(C.&) if $(p \& q) \epsilon \alpha$, then $p \epsilon \alpha$ and $q \epsilon \alpha$;

(C.v) if $(p \vee q) \epsilon \alpha$, then $p \epsilon$ or $q \epsilon \alpha$ (or both;

(C.∃) if $(\exists x)p \epsilon \alpha$, then $p(i/x) \epsilon \alpha$ for some free variable 'i';

(C.U) if $(x)p \epsilon \alpha$, and if the free variable 'i' occurs in the formulae of α, then $p(i/x) \epsilon \alpha$.

A semantics for deontic logic involves more than one model set. Obligation and permission involve possible worlds ("Kingdoms of Ends") related to the actual world. These are model systems, related sets of model sets. More precisely a *model system* M for nonrelativized deontic logic is an ordered pair $\langle \Omega, R \rangle$, where Ω is a set of model sets and R is a two-place relation on Ω, i.e., $R \subseteq \Omega^2$, such that:

(C.ϵ) every member μ of Ω is a model set;

(C.O*) if $Op \epsilon \mu \epsilon \Omega$, $\mu^* \epsilon \Omega$, and $R(\mu^*, \mu)$, then $p \epsilon \mu^*$;

(C.O) rest if $Op \epsilon \mu^*$ and
$R(\mu^*, \mu)$ for some $\mu \epsilon \Omega$, then
$p \epsilon \mu^*$;

(C.OO*) if $Op \epsilon \mu \epsilon \Omega$, $\mu^* \epsilon \Omega$, and $R(\mu^*, \mu)$, then $Op \epsilon \mu^*$;

(C.P*) if $Pp \epsilon \mu\Omega$, then $p \epsilon \mu^*$ for some μ^* such that $R(\mu^*, \mu)$;

(C.o*) if $Op \epsilon \mu \epsilon \Omega$, then $p \epsilon \mu^*$ for some μ^* such that $R(\mu^*, \mu)$.

Ω is interpreted as a set of deontically related possible worlds. Ω^2 is the Cartesian product $(\Omega \times \Omega)$ of Ω. R is "the relation of alternativeness." '$R(\mu^*, \mu)$' is to be read as 'μ^* is an alternative to μ' or 'μ^* is perfect relative to μ'.

Our basic semantic concepts follow. Any formula p is *satisfiable,* if and only if it is a member of some model set of some model system. Any formula p is *valid,* if and only if ~p is not satisfiable. For any formulae p and q, q is a *logical consequence* of p, i.e., q is logically implied by p, if and only if (p ⊃ q) is valid.

Various deontic formulae can be proved valid by using these (C) conditions (Cf. Hintikka [8, pp. 193–94, 197–98]). Proofs of the validity of p start from the assumption that ~p is *satisfiable,* that is, is a member of some model set in some model system. Via the (C) conditions, one then deduces a contradiction from this assumption. The contradiction shows that ~p was really not satisfiable, and that, thus, p is valid. The semantics for relativized deontic modalities begins with the above nonrelativized semantics.

Above the letters i, j, k, . . . are arbitrary free act-individual variables. Now add the letters a, b, c, . . . as arbitrary free individual *person* variables. Let U be the class of all free individual-person variables that occur in the formulae of a model set μ. Let A be an arbitrary subset of U. I will relativize the deontic operators O and P to subsets of U. O_A and P_A are deontic operators relativized to A, where $A \subseteq U$. 'P_A p' is to be read 'p is permissible for the members of A' or in short, 'p is permissible for A'. 'O_A p' is to be read 'p is obligatory for the members of A' or 'p is obligatory for A'. 'F_A p' is to be read 'p is forbidden for members of A' or 'p is forbidden for A'. I will define O_A and F_A in terms of P_A:

(D.3) $O_A p \equiv \sim P_A \sim p$;

(D.4) $F_A p \equiv O_A \sim p$.

If A is a unit set, e.g., $A = \{a\}$ where $a \in U$, then O and P are relativized to a single individual. $O_{\{a\}}$ and $P_{\{a\}}$ are then read as 'it is obligatory for a' and 'it is permissible for a', respectively.

With "impersonal" deontic modalities, a model set μ is a 'possible world', and Ω is a related set of possible worlds. Obligation in μ means, among other things, truth in all alternates μ^*, μ^{**}, etc. to μ. Permissibility in μ means truth in some alternative μ^* to μ. Relativized modalities are analogous. If $P_{\{a\}}\ p \in \mu \in \Omega$, there is an "alternative" μ^* to μ such that $p \in \mu^*$. In the present case, however, μ^* cannot be just any arbitrary alternative to μ. It must be an alternative with respect to $\{a\}$ or, in short, an $\{a\}$-alternative. Similarly, for obligation. If deontic operators are relativized to individuals or sets of individuals, the relation of alternativeness must be relativized in the same way. Alternativeness with respect to $\{a\}$ is expressed by '$R_{\{a\}}$'. We will read '$R_{\{a\}}(\mu^*, \mu)$' as 'μ^* is an alternative to μ with respect to $\{a\}$' or 'μ^* is an $\{a\}$-alternative to μ'.

Let's generalize. For any subset A_i of U, let '$R_{Ai}(\mu^*, \mu)$' mean that 'μ^* is an A_i-alternative to μ'. A *relativized model system* M_{rel} for relativized deontic logic is an ordered n-tuple $\langle \Omega, R_{A_1}, R_{A_2}, \ldots R_{A_{n-2}}, R \rangle$. Ω, as before, is a set of (related) model sets. R satisfies our previous, nonrelativized (C) conditions. n is some natural number. For any i in the interval $1 \leqslant i \leqslant n-2$, the R_{A_i} are each two-place relations on Ω, i.e., each $R_{A_i} \subseteq \Omega,^2$ such that for no i is $R_{A_{i+1}} \subseteq R_{A_i}$. Further, the R_{A_i} fulfill the following conditions:

($C.O_{A_i}{}^*$) if $O_{A_i}\, p \in \mu \in \Omega$, $\mu^* \in \Omega$, and $R_{A_i}(\mu^*, \mu)$, then $p \in \mu^*$;

($C.O_{A_i}$) rest if $O_{A_i}\, p \in \mu^* \in \Omega$ and $R_{A_i}(\mu^*, \mu)$ for some $\mu \in \Omega$, then $p \in \mu^*$;

($C.OO_{A_i}{}^*$) if $O_{A_i}\, p \in \mu \in \Omega$, $\mu^* \in \Omega$, and $R_{A_i}(\mu^*, \mu)$, then $O_{A_i}\, p \in \mu^*$;

($C.P_{A_i}{}^*$) if $P_{A_i}\, p \in \mu \in \Omega$, then $p \in \mu^*$ for some $\mu^* \in \Omega$ such that $R_{A_i}(\mu^*, \mu)$;

($C.o_{A_i}{}^*$) if $O_{A_i}\, p \in \mu \in \Omega$, then $p \in \mu^*$ for some $\mu^* \in \Omega$ such that $R_{A_i}(\mu^*, \mu)$.

Intuitively, n is two plus the number of subsets to which a given formula is relativized. For example, in deciding whether '$(Pp \,\&\, O_B\, p) \supset O_A\, p$' is valid for $A \subseteq B$, we would consider relativized model systems that are ordered 4-tuples and of form $\langle \Omega, R_A, R_B, R \rangle$. So far, these conditions are all analogous to those for nonrelativized deontic modalities. There will, however, be other conditions.

Relativized model systems involve R as well as the R_{A_i}. A question, therefore, arises. It can be expressed in several ways: What is the relation between A-alternatives to μ and alternatives to μ? What is the relation between relativized alternatives to μ and regular, nonrelativized alternatives to μ? What is the relation between R_A and R? The answers do not follow from the above (C) conditions. We should thus adopt some "bridge principle(s)" connecting relativized and nonrelativized deontic modalities. One of my candidates is the following (for any subset A of U):

(C. d-bridge) If μ^* is an A-alternative to μ in Ω, then μ^* is an (nonrelativized) alternative to μ.

(C. d-bridge) says that if $R_A(\mu^*, \mu)$, then $R(\mu^*, \mu)$, i.e., $R_A \subseteq R \subseteq \Omega^2$. (C. d-bridge) clearly imposes an additional restriction on R and on the R_{A_i}.

Given (C. d-bridge), the following are valid for any $A \subseteq U$:

(2.1) $O_p \supset O_A p$;

(2.2) $P_A p \supset P_p$.

Their respective converses

(2.3) $O_A p \supset Op$,

(2.4) $P_p \supset P_A p$

are not valid. Their nonvalidity shows that relativized deontic modalities are not "reducible" to nonrelativized deontic modalities.

To see that (2.1) is valid, assume that its negation

(2.5) $O_p \;\&\; P_A \sim p$,

is satisfiable, and that it belongs to some model set μ in some (relativized) model system,

(2.6) $(O_p \;\&\; P_A \sim p) \in \mu \in \Omega$.

From (2.6), (C. &), and (C.O*), there exists a $\mu^* \in \Omega$, such that $R(\mu^*, \mu)$ and

(2.7) $p \in \mu^*$.

Further, from (2.6), (C. &), and (C. P_{A_i}*), there exists a $\mu^{**} \in \Omega$ such that $R_A(\mu^{**}, \mu)$ and

(2.8) $\sim p \in \mu^{**}$.

Since $R_A(\mu^{**}, \mu)$, it follows that $R(\mu^{**}, \mu)$ in virtue of (C. d-bridge). From $R(\mu^{**}, \mu)$, (2.6), (C. &), and (C.O*), it follows that

(2.9) $p \in \mu^{**}$.

But (2.8) and (2.9) violate (C.~). Hence, (2.5) is not satisfiable. Its negation, (2.1), is therefore valid. (2.2) is also valid, given

(D.1), (D.3), and the validity of (2.1). (2.3) is not valid because its negation

(2.10) $O_A p$ & $P{\sim}p$

is satisfiable. (2.10) is satisfied by the relativized model system $M\ rel_o = \langle \Omega_o, R_A, R \rangle$. Ω_o consists of the two model sets

(2.11) $\mu = \{(O_A p \ \& \ P{\sim}p), O_A p, P{\sim}p\}$

and

(2.12) $\mu^* = \{O_A p, p\}$,

such that $R_A\ (\mu^*, \mu)$, along with the model set

(2.13) $\mu^{**} = \{{\sim}p\}$,

such that $R(\mu^{**}, \mu)$. $M\ rel_o$ satisfies all of the above (C) conditions. (2.10) is thus satisfiable. Its negation (2.3) is, therefore, not valid. (2.4) is also not valid, given (D.1), (D.3) and the nonvalidity of (2.3).

Given the foregoing, it is also plausible to assume that all non-relativized alternatives to μ are U-alternatives to μ:

(C.non-U) if μ^* is an alternative to μ in U, then μ^* is a U-alternative to μ.

(C.non-U) says that if $R\ (\mu^*, \mu)$, then $R_U\ (\mu^*, \mu)$, i.e., $R \subseteq R_U \subseteq \Omega^2$. Given (C. non-U) and (C. d-bridge),

(2.14) $O_p \equiv O_U p$

and

(2.15) $P_p \equiv P_U p$

are valid. The "impersonal" deontic concepts O and P need not therefore be primitive. They can be defined by (2.14) and (2.15). Alternately, (C. d-bridge) and (C. non-U) follow from (2.14), (2.15), and, for any subset A of U,

(C.A-U) if μ^* is an A-alternative to μ in Ω, then μ^* is a U-alternative to μ.

(C.A-U) says that if $R_A\ (\mu^*, \mu)$, then $R_U (\mu^*, \mu)$ i.e., $R_A \subseteq R_U \subseteq \Omega^2$. We can generalize (C. A-U). For any two subsets A, B, of U,

(C. d$\subseteq$) if $A \subseteq B \subseteq U$, then whenever μ^* is an A-alternative to μ in Ω, μ^* is a B-alternative to μ.

(C.d$\subseteq$) says that if $A \subseteq B$, then $R_A \subseteq R_B$. (C. d$\subseteq$) relates deontic concepts relativized to different sets of individuals. Considering (2.14) and (2.15) as definitions, (C. d-bridge), (C. non-U), and C. A-U) follow from (C. d$\subseteq$).

In light of (C. d$\subseteq$) the following schemata are valid if $A \subseteq B$:

(2.16) $O_B p \supset O_A p$,

(2.17) $P_A p \supset P_B p$.

(2.16) accords well with our intuitions. The following are some examples: What is obligatory for everyone is also obligatory for me. What is obligatory for Americans is also obligatory for Chicagoans. Remembering definitions (D.3) and (D.4), (2.17) also accords well with out moral intuitions. The following are some examples: What is permissible for me is not what is forbidden to everyone, that is, nothing is permitted to someone which is forbidden to everyone, or a version of "no one is above the law." What is permissible for Chicagoans is not what is forbidden to Americans, that is, nothing is permitted to Chicagoans which is forbidden to Americans.

Consider next "collective obligations." First note that

(2.18) $O_{\{a,b\}} p \supset (O_{\{a\}} p \; \& \; O_{\{b\}} p)$

is valid. Both $\{a\} \subseteq \{a,b\}$, and $\{b\} \subseteq \{a,b\}$. (2.18) is therefore valid because (2.16) is valid. Does this formal result accord with our informal intuitions concerning "collective obligations"? Collective obligations result when individuals collaborate. An example due to Rescher [11, p. 69] is

(a) John and Tom ought to move the table across the room together.

(a) expresses a collective obligation presumably because

(b) John ought to move the table across the room.

does not follow from (a). Such examples presumably falsify (2.18). One might thus claim that the theory presented here conflicts with our ordinary intuitions concerning collective obligations.[1]

Actually, however, there is no conflict. Such examples, correctly formalized, do not falsify (2.18). The invalidity of inferences from (a) to (b) is not peculiar to the deontic modalities. Such inferences are analogous to fallacies of composition and division, and to fallacies involving adverbial modification.[2] A fallacy of composition and division is the inference from

(c) John and Tom weigh four hundred pounds together

to

(d) John weighs four hundred pounds.

And a fallacy involving adverbial modification is the inference from

(e) this is a counterfeit coin

to

(f) this is a coin.

There are two ways to correctly formalize the invalid inference from (a) to (b). Neither is an instance of the valid (2.18). The first way is

(2.19) $O_{(a,b)}\ p \supset O_{\{a\}}\ p.$

Using the Wiener-Kuratowski definition of ordered pair, (2.19) becomes

(2.20) $O_{\{\{a\},\ \{a,b\}\}}\ p \supset O_{\{a\}}\ p.$

Clearly, (2.20) is not an instance of valid (2.18). Indeed (2.20) is invalid within the theory presented here. A heavy table and delinquent partner make (2.20) false and show that the move from (a) to (b) is invalid. For the same logical reason—which is independent of the nature of deontic modalities—the inference from (c) to (d) is invalid.

A second way to formalize the invalid inference from (a) to (b) is

(2.21) $O_{\{a\}}\ p \supset O_{\{a\}}\ q.$

This is because it can be expressed as the inference from

(a′) John ought to move the table across the room together with Tom to

(b′) John ought to move the table across the room.

This inference is invalid for the same reason that the inference from (e) to (f) is invalid. The invalid (2.21) clearly is not an instance of the valid (2.18).

Collective obligations do not, therefore, conflict with the theory presented here. The theory accords well with the intuitive concepts "obligatory for" and "permissible for."

III

The following axiom system plausibly corresponds to the above semantical theory. Rules of inference are those for the propositional calculus and for quantification theory, along with

(R. 1) from $p \equiv q$ to infer $P_A p \equiv P_A q$, for any $A \subseteq U$.

An infinite number of axioms are introduced by: (i) axiom schemata for the propositional calculus and for quantification theory and (ii) for any $A \subseteq U$, the deontic schemata:

(A. 1) $O_A p \supset \sim O_A \sim p$;

(A. 2) $O_A (p \& q) \equiv O_A p \& O_A q$;

(A. 3) $O_A (p v \sim p)$;

(A. 4) $O_A p \supset O_A O_A p$;

(A. 5) $O_A (O_A p \supset p)$;

(A. 6) $O_A p \supset O_B p$, if $B \subseteq A$.

The operator P_A is defined as $\sim O_A \sim$. (I will not here formulate natural deduction rules for relativized deontic modalities; cf. Fitch's elegant formulation of such rules for nonrelativized deontic modalities [2].)

IV

The theory presented here illuminates supererogation: the saint and the hero. There is no theoretical niche for acts of supererogation within most ethical theories. Most ethical theories recognize only three types of moral actions: the obligatory (for all); the permissible (for all), but not obligatory (for all); the forbidden (for all). Acts of supererogation do not fall within this "trichotomy of duties, indifferent actions, and wrong-doing" [12, p. 73], but they have great moral significance. Most ethical theories are deficient because they do not accommodate supererogatory actions.[3] An important task is the construction of a general theory which accommodates acts of supererogation.[4] The theory presented here takes a first step in this direction.

What is supererogation? Saints and heroes perform supererogatory actions. These actions are "far beyond" the "basic" or "rock-bottom duties for all."[5] Heroic and saintly deeds provide the most noteworthy examples of morally significant actions that lie outside the usual trichotomy. There are other, less exalted, examples. All members of a club have obligations that are a condition of membership.[6] But there may be a few whose devotion and loyal service far exceeds what is required of all. Such individuals are to their club just as saints and heroes are to mankind.

There is an important distinction between basic obligations, binding on all, and "the higher flights of morality of which saintliness and heroism are outstanding examples" [12, p. 70]. Urmson's remarks are suggestive:

> . . . we may look upon our duties as basic requirements to be universally demanded as providing the only tolerable basis of social life. The higher flights of morality can then be regarded as more positive contributions that go beyond what is universally to be exacted; but while not exacted publicly they are clearly equally pressing *in foro interno* on those who are not content merely to avoid the intolerable (p. 73).

Like traditional moral theory, nonrelativized deontic logic lacks a theoretical niche for these distinctions. There is no way to say that an action is obligatory for a person or group, but not for everyone. That is, neither

(4.1) $O_{\{a\}}\, p \;\&\; \sim O_{\mathbf{U}}\, p$

nor

(4.2) $O_{A}\, p \;\&\; \sim O_{\mathbf{U}}\, p$

is *expressible* within nonrelativized deontic logic. They do not there have a theoretical niche. (4.1) and (4.2) are, however, expressible within the theory for relativized deontic modalities presented here. They have a theoretical niche here which can be filled. That is, within the present theory, (4.1) and (4.2) are satisfiable since, by (2.3) and (2.14), neither

(4.3) $O_{\{a\}}\, p \supset O_U p$

nor

(4.4) $O_A\, p \supset O_U p$

is valid.

Also note that, because of (2.16),

(4.5) $O_U p \supset O_{\{a\}}\, p$

is valid. If 'a' names a saint or hero, then (4.5) says that what is obligatory for all is also obligatory for a saint or hero. The intuitive correctness of this assertion is sytematically preserved by the validity of (4.5). Interpreted thus, (4.3) then says that what is obligatory for a saint or hero is also obligatory for all. The intuitive incorrectness of this assertion is systematically preserved by the nonvalidity of (4.3).

Consider Alexander the Great, a hero, cutting the Gordian knot. Let 'Op' stand for 'The Gordian knot ought to be cut.' Because he is unlike other men, Op obtains for Alexander (a), but not for others. Within the present theory for relativized deontic modalities, we can express this by $O_{\{a\}}\, p\ \&\ {\sim}O_U p$. Whatever else it may be, the supererogatory act, p, of some saint or hero, a, is obligatory for a, but not for everyone.[7] The present theory for relativized deontic modalities takes a large first step, therefore, in solving the problem of supererogation.

References

[1] Chisholm, R. M. 1963. Contrary-to-duty imperatives and deontic logic. *Analysis* 24: 33–36.

[2] Fitch, F. B. 1966. Natural deduction rules for obligation. *American philosophical quarterly* 3: 27–38.

[3] Føllesdal, D., and Hilpinen, R. 1971. Deontic logic: an introduction. In *Deontic logic: introductory and systematic readings,* R. Hilpinen, ed., pp. 1–35. Dordrecht, Holland: D. Reidel.

[4] Hansson, B. 1971. An analysis of some deontic logics. In *Deontic logic: introductory and systematic readings,* R. Hilpinen, ed., pp. 121–47. Dordrecht, Holland: D. Reidel.
[5] Hilpinen, R. 1969. An analysis of relativized modalities. In *Philosophical logic,* J. W. Davis, D. J. Hockney, and W. K. Wilson, eds., pp. 181–93. Dordrecht, Holland: D. Reidel.
[6] ——. 1973. On the semantics of personal directives. *Ajatus* 35.
[7] Hintikka, J. 1969. Epistemic logic and the methods of philosophical analysis. In *Models for modalities,* ed. J. Hintikka, pp. 3–19. Dordrecht, Holland: D. Reidel.
[8] ——. 1969. Deontic logic and its philosophical morals. In *Models for modalities,* ed. J. Hintikka, pp. 184–214. Dordrecht, Holland: D. Reidel.
[9] ——. 1971. Some main problems of deontic logic. In *Deontic logic: introductory and systematic readings.* R. Hilpinen, ed., pp. 59–104. Dordrecht, Holland: D. Reidel.
[10] Hohfeld, W. N. 1919. *Fundamental legal conceptions as applied in judicial reasoning,* ed. W. W. Cook. New Haven: Yale University Press.
[11] Rescher, N. 1966. *The logic of commands.* New York: Dover.
[12] Urmson, J. O. 1969. Saints and heroes. In *Moral concepts,* J. Feinberg, ed., pp. 60–73. London: Oxford University Press.

16: DAVID HARRAH

A System for Erotetic Sentences

In recent years 'erotetic logic' has been used to refer to the logic of questions. The main point of this paper is to outline a wider conception of the erotetic and to present a set of concepts and a system of derivation that can serve as a basis for the logic of erotetic sentences. Roughly speaking, in the logic of questions the main concept is that of wanted reply; in our wider conception of the erotetic the main concept is that of relevant reply. What seems to be new and peculiar about the theory presented here, aside from the generality of its erotetic concepts, is the routine of reply. The user of this system can reply directly to simple erotetic sentences; to reply to a compound, he uses the derivation system to extract simples from the compound, and then he replies to these simples. It will become clear incidentally that in a certain sense our logic of erotetic sentences reduces to the logic of declarative sentences, in line with a view suggested by Frederic B. Fitch.[1]

To fix ideas, we shall describe an erotetic logic for a particular formal language L. This language has a d-part ('d' for 'declarative') and an e-part ('e' for 'erotetic'), and perhaps other parts as well. The alphabet of L is finite, and each expression of L is a finite string of letters of the alphabet. The d-part of L is itself a standard language with first-order predicate logic, identity, and perhaps descriptions; as will be evident below, the e-part is an extension of the d-part.

For the d-part, L has denumerably many individual variables x, y, z, the usual connectives (–, &, V, ⊃, ≡), quantifiers (U, E), identity (=), and possibly a description operator. It has some constants of the usual sorts, including n-ary predicates, individual constants, and n-ary functors. *Term* and *d-wff* (declarative well-formed formula) are defined in one of the usual ways, depending on the theory of descriptions adopted. Freedom and bondage of variables, and proper substitution of terms for variables, are as usual; a d-sentence is a d-wff with no free variables. The d-part of L is then given the usual model-theoretic semantics.

For the d-wffs of L we choose a logic that is semantically complete in the standard sense. It would be possible to choose a natural deduction system for this, but it will simplify our exposition if we assume below that the d-logic of L is an axiomatic system having just the rules MP (modus ponens) and Gen (Generalization), and an appropriate, effectively specified set of d-wffs chosen as d-axioms. A *d-proof* is a finite, nonempty sequence Z of d-wffs such that, for every member F of Z, either F is a d-axiom or F comes from one or two preceding members of Z by Gen or MP.

The main application of erotetic logic is to communication situations, including question-and-answer situations. In these situations it is very convenient if the language L can refer to its own expressions and various set-theoretic constructions on them; with this capability, the users of L can talk to each other both about the primary subject matter of L (whatever it is) and about the messages they send to each other. Even apart from talking about messages, and apart from communication applications in general, a language with set-theoretic and syntax-theoretic capability will have an erotetic logic that is essentially enriched; an example will be noted below.

Accordingly, let us assume that L has constants that can be interpreted as expressing the basic notions of set theory and syntax theory, and that we have an effective set W* of d-wffs that contains axioms sufficient for Zermelo-Fraenkel set theory and Tarski's theory of syntax. Interpretations of L that assign the appropriate meanings to these constants and make the axioms in W* true are the normal or intended interpretations.

A *normal d-derivation* from S is a finite, nonempty sequence Z of d-wffs such that, for every member F of Z, either: (1) F is a d-axiom, or a member of W*, or a member of S; (2) F comes from two preceding members of Z by MP; or (3) F comes from one preceding member G of Z by Gen, where G is the last member of a subsequence Z′ of Z such that Z′ is a d-proof.

We say that F is *normally d-derivable from* S just in case F is the last member of some normal d-derivation from S (notation: $S\vdash^* F$, and $G\vdash^* F$ in case S consists of just the wff G, and $\vdash^* F$ in case S is empty). Also, S is *normally d-derivation consistent* just in case there is no F such that both S⊢*F and S⊢*–F. It should be clear in our discussion below which parts of our erotetic logic depend only on the standard grammatical and logical framework of L, and which parts depend on the special set-theoretic and syntax-theoretic content of L.[2]

To construct the e-part of L, we begin with a denumerable set of ae-operators ('ae' for 'atomic erotetic'). Each of these is an expression of L, is effectively recognizable as being an ae-operator, and is not identical with any expression that occurs in any d-wff of L. In our discussion of examples below we shall be concerned with seven particular ae-operators. Let us refer to these via the metalinguistic signs: $?^{w}$, $?^{1x}$, $?^{cx}$, $?^{y}$, $?^{z}$, !, †.

An *ae-wff* is an expression of L of the form $X(Y_1, \ldots, Y_n)$, where $n \geqslant 1$, X is either an ae-operator or an ae-operator followed by a string of distinct variables, and each Y_i is either a term or a d-wff.

Next, for certain types of ae-wff and d-wff, various notions of content and reply are defined. These notions and their interconnections can be summarized as follows. If F is either an ae-wff or a d-wff, then (1) there are some indicated replies to F; (2) every indicated reply to F is a d-wff; (3) every wanted reply to F is an indicated reply to F; (4) the assertive core of F is a d-wff G such that (a) G is implied by every indicated reply to F, (b) every d-wff that is implied by all the indicated replies to F is implied by G, and (c) given F, G is effectively recognizable as the assertive core of F; (5) the corrective reply to F is the negation of the assertive core of F; (6) the direct replies to F are the indicated replies plus the corrective reply; (7) the full replies to F are the d-wffs that imply direct replies; (8) the partial replies to F are the d-wffs that are implied by direct replies; (9) the relevant replies to F are the full replies plus the partial replies.

Some examples of what is allowed for in this scheme are the following:

(1) *Atomic whether questions.* Expressions of the form $?^{w}(F_1, \ldots, F_n)$, where $n \geqslant 2$, and $F_1, \ldots, F_n$ are distinct d-wffs, can be used to ask whether questions of a certain kind. The assertive core here would be $(F_1 \vee \ldots \vee F_n)$, the indicated replies would be $F_1, \ldots, F_n$, the wanted replies would be $F_1, \ldots, F_n$, and the corrective reply would be $-(F_1 \vee \ldots \vee F_n)$.

(2) *Atomic one-example questions.* Expressions of the form

$?^{1x} x_1 \ldots x_n(F)$, where F is a d-wff, can be used to request a single example of an n-tuple of things that stand in the relation expressed by F. The assertive core would be $Ex_1 \ldots Ex_n F$. The indicated replies would be the d-wffs that come from F by proper substitution of terms for $x_1, \ldots, x_n$; and the wanted replies would be the same as the indicated replies.

(3) *Atomic complete-list questions.* Expressions of the form $?^{cx} x_1 \ldots x_n(F)$ can be used to request a complete list of the n-tuples of things that stand in the relation F. The assertive core would be a d-wff saying that there is a finite set S such that Z is in S if and only if Z is an n-tuple of things that stand in the relation F. This assertion can be expressed in L because L has the resources of set theory. The indicated replies would have the obvious form, and the wanted replies would be the same as the indicated replies.

(4) *Atomic why questions.* Expressions of the form $?^{y}(F)$, where F is a d-wff, can be used to ask "Why is it the case that F?" The assertive core here would be F. Each indicated reply would be a conjunction whose conjuncts are the items in a d-derivation of F from some premise that "explains" F. Those who believe that we do not yet have a good theory of what counts as a good explanation may choose to leave the notion of wanted reply undefined for this type of question.

(5) *Atomic who questions.* Expressions of the form $?^{z}(t)$, where t is a term, can be used to express "Who is t?" or "Tell me about t." The assertive core would be $Ex(x = t)$, where x is not in t. As with the why questions above, we could define indicated reply in a very general way and leave wanted reply undefined.

(6) *Atomic imperatives.* Expressions of the form !(F, G), where F and G are distinct d-wffs, can be used to express a certain kind of directive or imperative. The assertive core would be $(F \vee G)$, the indicated replies would be F and G, but the only wanted reply would be F. Roughly, F is the desired state of affairs, and G is the threatened alternative.[3]

(7) *Atomic pronouncements.* Expressions of the form †(F), where F is a d-wff, can be used to express a certain kind of proclamation, pronouncement, or proposal, as in "I hereby declare that F." The assertive core here would be F, and F is both the one and only indicated reply and the one and only wanted reply. (The speaker wants to know that the hearer agrees with him.)

(8) *Atomic assertions.* For any d-wff F, we let F itself be its own

assertive core, and we let F be the one and only indicated reply, but there is no wanted reply. Thus, to an assertion F no reply is called for; but some replies are relevant, including both F and -F.

Problem for research: Determine whether all of the atomic erotetic sentences of natural language can be rendered in the form of our ae-wffs as defined above.

Our general notion of erotetic wff is defined recursively: (1) every ae-wff is an e-wff; (2) every d-wff is an e-wff. (3) if F and G are e-wffs, and x is a variable, then the following are e-wffs: (F & G), (F V G), UxF, ExF; (4) if F is a d-wff, and G is an e-wff, then (F ⊃ G) is an e-wff; (5) nothing is an e-wff unless its being so follows from (1)–(4).

For d-wffs we have assumed the usual notions of bondage, freedom, proper substitution, etc. We extend these notions to e-wffs by treating the X in an ae-wff $X(Y_1, \ldots, Y_n)$ as an operator that binds all the variables occurring in X. For example, in the ae-wff $?^x$xy(F), the variables x and y are bound, but any variable distinct from x and y that occurs free in F is free.

For any e-wff F, G is the *assertive core* of F if and only if G is like F, except that wherever F contains an ae-wff H, G contains the assertive core of H. To refer to the assertive core of F we write: Acore(F). For any set S of e-wffs, to refer to the set of assertive cores of members of S we write: Acore(S). An e-wff F is *assertively valid* just in case Acore(F) is valid.

We add one further assumption about ae-wffs. We assume that assertive cores are assigned to ae-wffs in such a way that the following holds for all e-wffs: If Ga comes from Gx by proper substitution of the term a for x, then there is a d-wff F such that F is d-equivalent to Acore(Gx), and Acore(Ga) comes from F by proper substitution of a for x. (This assumption is needed for T13–17 below.)

Next we develop our system of erotetic derivation. To see what is wanted here, consider the case where S is a set of e-wffs, F is an e-wff, and Acore(S) implies Acore(F). Eventually, for certain purposes, we shall want a system that can generate all the assertively valid e-wffs and that can validate the inference from S to F in cases like the one just described. Indeed, much work has already been done by Belnap, Åqvist, and others to develop the model-theoretic background for systems of this sort. What we want for our present purposes, however, is different. Roughly speaking, what we need here is a system such that when a sender conveys a set S of e-wffs to a receiver, the receiver can use the system to derive from S all (and only) the ae-wffs to which he should reply. For cases like the one considered above we do want Acore(F) to be normally d-derivable from Acore(S), but we do not want F to be erotetically derivable from S unless F is one of the wffs to which the sender wants or expects a reply.

Our rules of e-derivaton are indicated by eleven schemata. In all cases F, F′, F″ are e-wffs, and G is a d-wff. In each case we say that the wff to the right of the arrow is *immediately e-derivable from* the wff or wffs to the left of the arrow: (1) (F & F′) → F; (2) (F & F′) → (F′ & F); (3) ((F & F′) & F″) → (F & (F′ & F″)); (4) (F V F) → F; (5) (F V F′) → (F′ V F); (6) ((F V F′) V F″) → (F V (F′ V F″)); (7) –G, (G V F) → F; (8) G, (G ⊃ F) → F; (9) UxF → F′, where F′ comes from F by proper substitution of some term for the variable x; (10) ExF → (Acore(F′) ⊃ F′), where F′ comes from F by proper substitution of some term for x; (11) F → Acore(F).

It seems appropriate to call (8) the MP rule for e-wffs, (9) the UI rule, (10) the EI rule, and (11) the A rule. We shall discuss the incompleteness of this set of rules later, after we note the use of this derivation system in our system of replies.

We say that Z is an *e-derivation from* S if and only if Z is a finite, nonempty sequence of e-wffs such that, for every member F of Z, either (1) F is a d-axiom, a member of W*, or a member of S; (2) F is immediately e-derivable from preceding members of Z; or, (3) F is a d-wff, and F comes by Gen from a preceding member G of Z such that G is the last member of a subsequence Z′ of Z that consists of d-wffs and is a d-proof.

We call Z an *ec-derivation from* S if and only if Z is a finite, nonempty sequence of e-wffs such that, for every member F of Z, either (1)–(3) as above, or (4) there is a G such that (a) (F V G) is an e-wff but not a d-wff, (b) either (F V G) or (G V F) precedes F in Z, and (c) if G occurs in Z, then either G is a member of S, or G is immediately e-derivable from members of Z that occur in Z before the first occurrence of G in Z.

A wff F is *e-derivable from* S if and only if F is the last member of some e-derivation from S. Notation: $S \vdash^{e} F$, and $G \vdash^{e} F$ in case S consists of just the wff G, and $\vdash^{e} F$ in case S is empty. Similar terminology and notation will be used with our ec-concepts.

In an ec-derivation we use the rules of e-derivation and a new rule, which may be called the C rule, as indicated in clause (4) of the definition of ec-derivation. The letter 'C' is to suggest choice. In effect, if one has an e-wff (F V G), he may freely choose one of the disjuncts and add it to the derivation. He cannot freely choose both disjuncts. If he chooses F, he can have G only if he can get G by other means.

The *assertive content* of S, or Acont(S), is the set of all d-wffs F such that $S \vdash^{e} F$. S *assertively contains* S′ if and only if Acont(S) includes Acont(S′); and two sets are *assertively equivalent* if and only if they assertively contain each other. S is *assertively consistent* just in case Acont(S) is normally d-derivation consistent.

The following metatheorems express some important properties of these types of derivation:

T1. If S is effective, then there is an effective test for deciding, of a given finite sequence Z, whether or not Z is a normal d-derivation from S. Similarly for e-derivation and ec-derivation.

T2. There is no effective test for deciding, for arbitrary S and F, whether or not $S \vdash^{*} F$; similarly for $\vdash^{e}$ and $\vdash^{ec}$.

T3. If Z and Z′ are normal d-derivations from S, then so is ZZ′, where ZZ′ is the sequence consisting of first the members of Z in order and then the members of Z′ in order; similarly for e-derivation, but not for ec-derivation.

T4. If F and (F ⊃ G) are normally d-derivable from S, so is G; similarly for e-derivation, but not for ec-derivation.

T5. If $S \vdash^{*} F$ for every F in S′, and $S' \vdash^{*} G$, then $S \vdash^{*} G$; similarly for $\vdash^{e}$, but not $\vdash^{ec}$.

T6. If $\vdash^{e} F$ or $\vdash^{ec} F$, then F is a d-wff and $\vdash^{*} F$.

T7. Unless $\vdash^{*} F$, $S \vdash^{*} F$ if and only if there are $G_1, \ldots, G_n$ in S such that $(G_1 \mathbin{\&} \ldots \mathbin{\&} G_n) \vdash^{*} F$; the same does not hold for either $\vdash^{e}$ or $\vdash^{ec}$.

T8. If $S \cup \{F\} \vdash^{*} G$, then $S \vdash^{*} (F \supset G)$; the same does not hold for $\vdash^{e}$ or $\vdash^{ec}$.

T9. Assertive containment and assertive equivalence are reflexive and transitive.

T10. Normal d-derivation consistency is preserved under normal d-derivation, but not under e-derivation or ec-derivation. Assertive consistency is preserved under normal d-derivation and e-derivation, but not under ec-derivation.

T11. Suppose that S is assertively consistent. Then, not every d-wff is e-derivable from S, but it might be the case that every d-wff is ec-derivable from S. (Example: Let S contain $(F \vee ?^{y}(-F))$ for every d-wff F.)

T12. Suppose that $S \vdash^{ec} F$, and that F' is an ae-wff occurring in F. Then there is an e-wff G in S such that, for some G', G' occurs in G, and F' comes from G' by proper substitution of terms for variables. Corollary: If $S \vdash^{ec} F$, then every ae-operator occurring in F occurs in some member of S.

T13. If $F \vdash^{e} G$, then $\text{Acore}(F) \vdash^{e} \text{Acore}(G)$.

T14. If G is a d-wff, and $F \vdash^{e} G$, then $\text{Acore}(F) \vdash^{e} G$.

T15. For any set S of e-wffs, S is assertively equivalent to Acore (S).

T16. Assertive validity is preserved under e-derivation; in particular, if $F \vdash^{e} G$, then $\vdash^{*}(\text{Acore}(F) \supset \text{Acore}(G))$. The same does not hold for $\vdash^{ec}$.

T17. It is always safe to use e-wffs of the form $(\text{Acore}(F) \supset F)$. This is because the assertive content of such a wff consists of d-wffs G such that $\vdash^{*} G$.

Problem for research: Find reasonable restrictions to put on our C rule, such that we can still make choices, and ec-derivation will still be effective, but ec-derivations will be better behaved.

Let S and S' be any sets of expressions of L. Then, (1) S'' is a *framework for reply to* S *on the basis of* S' *via* Z if and only if Z is an ec-derivation from $S \cup S'$, and S'' is a set consisting of d- and ae-sentences that are members of Z. (2) the *erotetic content of* S *on the basis of* S' is the set of all sets S'' such that, for some Z, S'' is a framework for reply to S on the basis of S' via Z; (3) S''' is a *reply set for* S *on the basis of* S' if and only if there are S'' and Z such that S'' is a framework for reply to S on the basis of S' via Z, and S''' is a set of pairs $\langle X, Y \rangle$ such that X is a member of S'', and Y is a relevant reply to X.

In applications of these concepts, S is what the sender conveys to the receiver, S′ consists of some sentences that express some of the receiver's beliefs, Z is an ec-derivation that the receiver makes from S with the aid of S′, and S″ consists of some of the d- and ae-sentences in Z. Very roughly, S″ is an aspect of S to which the receiver constructs the reply S‴; the erotetic content of S consists of all the aspects of S to which the receiver may reply. Incidentally, a framework for reply to S may contain d-sentences that are valid but irrelevant to the receiver's interests. The receiver may choose to reply to such a sentence, but he need not; and the reply, if given, would at best be an indicated reply, for which the sender need not pay.

The above notion of a reply set is very general. A problem for research is to discriminate types of reply set, discover which types are important, and study these. Some important types of reply set are those that are complete in certain respects. Roughly, a complete ec-derivation would be one in which all the rules that could be applied were in fact applied; a complete framework for reply would be a framework that consisted of all the d- and ae-sentences in a complete ec-derivation; a complete direct reply set would be one that began with a complete framework S″ and gave direct replies to all the members of S″ (and analogously for complete wanted reply set, complete full reply set, and the like). The key to such an analysis, obviously, is a precise definition of complete ec-derivation. A complete ec-derivation must of course be an ec-derivation and hence finite; it seems reasonable to require also that complete ec-derivations be effectively recognizable as being complete. Given these requirements, it seems natural to begin with a definition like the following.

An ec-derivation Z is *complete for* S just in case Z is an ec-derivation from S, every e-wff in S is a member of Z, and (1) if (F & F′) is in Z, so is F; (2) if (F & F′) is in Z, so is (F′ & F), . . . (8) if G and (G ⊃ F) are in Z, so is F; (9) if UxF is in Z, so is each e-wff that comes from F by proper substitution of a term t for x, where t is a term occurring in some e-wff in S; (10) if ExF is in Z, so is (Acore(F′) ⊃ F′), where F′ comes from F by proper substitution of a term t for x, for some term t; (11) if F is in Z, so is Acore(F); (12) if (F ∨ G) is in Z and is not a d-wff, then either F or G is in Z.

It turns out that under this definition some complete ec-derivations are more complete than others. Consider first the case where S consists of ((F & F′) ⊃ G), where G is an e-sentence, and S′ consists of F and F′. It is possible to derive from S∪S′ first (F & F′) and then G. The receiver of S might be motivated and skilled enough to do this, but nothing in our definition of complete ec-derivaton forces him to do so, and hence G will belong to some complete ec-derivations but not to others. The sender of S could help matters by using (F ⊃ (F′ ⊃ G)) instead of ((F & F′) ⊃ G). More generally, some ways of formulating the members of S are more fruitful than others. There are well-known rules for "cleaning up" a set of wffs: by cancellation of double negations and vacuous quantifiers, and judicious rewriting of bound variables, for example. We might have incorporated these techniques in our rules of e-derivation and in our definition of complete ec-derivation, but this would be pointless unless the sender and receiver cooperate in their formulation of S and S′. In any case, no such set of rules will force the production of a unique derivation that is most complete and is completely free of unwanted items.

Consider the case where S consists of (ExFax ⊃ G), (ExFbx ⊃ H), and Ey$?^{1x}$x(Fyx), where G and H are ae-sentences, and S′ consists of Fac, Fbc, and Fcc. The receiver will begin Z by putting in all the members of S and S′, and then the assertive cores of these wffs. Now he could choose a new term d, derive (ExFdx ⊃ $?^{1x}$x(Fdx)), derive the assertive core of this, and then stop. Alternatively, or additionally, he could derive (ExFcx ⊃ $?^{1x}$x(Fcx)), then ExFcx, and then $?^{1x}$x(Fcx). Alternatively, or additionally, he could derive (ExFax ⊃ $?^{1x}$x(Fax)), ExFax, $?^{1x}$x(Fax), and G as well. Alternatively, or additionally, he could derive analogous wffs using b, arriving finally at H. Thus, there are some complete ec-derivations here with neither G nor H, some with just one, and some with both. It seems unreasonable to require all the possibly relevant uses of the EI rule. If, however, we require that this rule be used only once, which application should we choose?

At this point we may conjecture that the best way to proceed, or at least a good way, is the following: (1) keep our rules and definitions approximately as they are above, possibly adding to our rules of e-derivation a few "cleanup" rules; (2) define the *prime erotetic content* of S as the set of all complete frameworks for reply to S; (3) define the *erotetic focus* of S as the intersection of the prime erotetic content of S; (4) study how various strategies for constructing S can affect the erotetic focus, especially the size of the erotetic focus relative to the prime erotetic content; (5) determine whether, for certain purposes, some strategies are optimal.

It is appropriate to conclude this paper as we have, with a conjecture; for a conjecture, being a hybrid of statement, question, and suggestion, is the erotetic entity *par excellence.*

Notes

Chapter 2

1 For more details about the structure and function of categorial frameworks, see my *Categorial Frameworks* (Oxford: Basil Blackwell, 1970 and 1974).
2 See my "Russell's Critique of Leibniz," (forthcoming).
3 For details, see my "Material Necessity," *Kant-Studien* (vol. 64, 1973, No. 4, pp. 423–430).
4 See my "Rational Choice," *Proceedings of the Aristotelian Society*, supp. vol. 47 (1973): pp. 1–18.
5 For details, see ibid.
6 Ibid.
7 *Système nouveau de la nature*, §17.
8 *Critique of Pure Reason* B253–256.
9 *An Enquiry Concerning Human Understanding*, ed. L. A. Selby-Bigge, 2nd ed. (Oxford: Clarendon Press, 1902), pp. 89 ff.
10 While a discussion of this problem is here out of the question, it seems permissible to express sympathy with W. Stegmüller's general approach, especially his distinction between inductive arguments belonging to the sphere of factual thinking and inductive arguments belonging to the sphere of practical thinking. See, in particular, *Probleme und Resultate der Wissenschaftstheorie und Analytischen Philosophie* (Berlin: Springer, 1973), vol. 4.
11 See J. Burnet, *Greek Philosophy–Thales to Plato* (London: MacMillan, 1943), §92.

Chapter 3

1 F. B. Fitch, "Attribute and Class," *Philosophic Thought in France and the United States*, ed. Marvin Farber (Buffalo: University of Buffalo Publications, 1950), p. 552.
2 W. V. O. Quine, *From a Logical Point of View*, 2d rev. ed. (New York: Harper Torchbook, 1961), p. 139.
3 Richard Cartwright, "Identity and Substitutivity," *Identity and Individuation*, ed. M. Munitz (New York: New York University Press, 1971), pp. 119–33.
4 See, for example, Leonard Linsky, "Reference, Essentialism and Modality," *Journal of Philosophy* 67 (1969): 687–88.
5 Peter Geach and Max Black, *Translations from The Philosophical Writings of Gottlieb Frege* (Oxford: Blackwell, 1952), p. 46n.

6 Cartwright, "Identity and Substitutivity," p. 120.
7 K. S. Donnellan, "Reference and Definite Descriptions," *Philosophical Review* 75 (1966): 281–304.
8 S. Kripke, "Identity and Necessity," in *Identity and Individuation,* ed. M. Munitz (New York: New York University Press, 1971), pp. 135–64. Also, idem., "Naming and Necessity," in *Semantics of Natural Language,* ed. G. Harmon and D. Davidson (Dordrecht: Reidel, 1972). K. S. Donnellan, op. cit.
9 P. T. Geach, *Logic Matters* (Berkeley and Los Angeles: University of California Press, 1972), p. 155.
10 Kripke and Geach's account of how proper names can come to have the use they do raises questions about the nature of those objects about which we can say, without further analysis, that an identity relation holds. If the requirement on identity statements is acquaintance with the objects named, and that there is a naming episode supposes that we are dealing with concrete individuals publicly displayable, then the question arises as to whether expressions used to designate abstract objects, whatever the syntactic simplicity of those expressions, are being used as proper names. Positing an archtypal white queen (of chess) or a number is not at the same time positing an object that can be properly named in any ordinary way.

Chapter 4

1 I wish to acknowledge my debt to K. Lambert and R. H. Thomason for many discussions on philosophy of mathematics, and to Canada Council grant S71-0546 for financial support.
2 A. Tarski and R. L. Vaught, "Arithmetical Extensions of Relational Systems," *Compositio Mathematica* 13 (1957): 81–102
3 E. Beth, *Wijsbegeerte der Wiskunde* (Antwerp: N. V. Standaard, 1948), p. 296 (my translation).
4 Ibid., p. 297.
5 E. Beth, "Carnap's Views on the Advantages of Constructed Systems over Natural Languages in the Philosophy of Science," in P. A. Schilpp, ed., *The Philosophy of Rudolf Carnap* (Open Court: LaSalle, Illinois, 1963), pp. 478–79.
6 K. Gödel, "What is Cantor's Continuum Problem?" in P. Benacerraf and H. Putnam, eds., *Philosophy of Mathematics* (Prentice Hall: Englewood Cliffs, N.J., 1964), p. 272.
7 See L. Henkin, "Some Notes on Nominalism," *Journal of Symbolic Logic* 18 (1953): 19–29; W. Sellars, "Grammar and Existence: A Preface to Ontology," in his *Science, Perception and Reality* (Humanities Press: New York, 1963), pp. 247–81; B. van Fraassen, *Formal Semantics and Logic* (Macmillan: New York, 1971), ch. 4, sec. 9.
8 H. Putnam, *Philosophy of Logic* (Harper & Row: New York, 1971), pp. 12–13.
9 Beth, *Wijsbegeerte,* p. 330.

10 K. Gödel, "Russell's Mathematical Logic," in Benacerraf and Putnam, *Philosophy of Mathematics*, p. 220.
11 Putnam, *Philosophy of Logic*, ch. 8, and a recent lecture at the University of Toronto.

Chapter 5

1 See especially the author's *Events, Reference, and Logical Form* (Washington: The Catholic University of America Press, to appear).
2 The oversimplified handling of tense here is Reichenbachian in some respects only. See his *Elements of Symbolic Logic* (New York: Free Press, 1966), pp. 287–98; and my own *Logic, Language, and Metaphysics* (New York: New York University Press, 1971), pp. 75–86.
3 Cf. George Lakoff, *Linguistics and Natural Logic, Studies in Generative Semantics* No. 1 (Phonetics Laboratory, The University of Michigan: 1970), and Zellig Harris, "The Two Systems of Grammar: Report and Paraphrase," in his *Papers in Structural and Transformational Linguistics* (Dordrecht: Reidel, 1972).
4 Charles J. Fillmore, "Lexical Entries for Verbs," *Foundations of Language* 4 (1968): p. 382. Cf. also his "The Case for Case," in *Universals in Linguistic Theory*, ed. by E. Bach and R. Harms (New York: Holt, Rinehart, and Winston, 1968). See also, for example, John Lyons, *Introduction to Theoretical Linguistics* (Cambridge: Cambridge University Press, 1968), pp. 289 ff.
5 Cf. the author's *Belief, Existence, and Meaning* (New York: New York University Press, 1969), chapter 6.
6 Further details will be forthcoming in the author's *Pragmatics, Logic, and Linguistic Structure*, in preparation. See also "On Prepositional Protolinguistics," to appear in the *Festschrift* for Paul Lorenzen, ed. by Kuno Lorenz.

Chapter 6

1 Professor H. N. Castaneda has objected that classical propositions (which he accepts) are not individuals. One reason for denying that they are individuals is that, in all languages, the (semantical) category of propositional expressions is different from that of individual expressions. I don't find this reason convincing by itself, because we can name propositions with individual expressions even though we express them with sentences. But even if classical propositions are something other than individuals, they deserve to be called quasi-individuals, for they have properties and are related to one another. They are also related to ordinary individuals. (I believe that, historically, talk about propositions has been modeled on talk about individuals.)
2 Both acts of omission and acts of commission can be done for a purpose. In this paper, however, I shall consider only acts of commission–acts that consist in doing something rather that in not doing it.

3 G. E. M. Anscombe, in *Intention* (Oxford: Basil Blackwell, 1958) also concludes that acts done on purpose for no purpose are not very important. She writes that

> the concept of voluntary or intentional action would not exist, if the question 'Why?', with answers that give reasons for acting, did not. Given that it does exist, the cases where the answer is 'For no particular reason', etc. can occur; but their interest is slight; and it must not be supposed that because they can occur that answer would everywhere be intelligible, or that it could be the only answer ever given (p. 34).

4 Recall how Socrates, in the *Phaedo,* ridicules the kind of explanation that Anaxagoras would give for Socrates' being in prison. What is wrong with an Anaxagorean explanation is that it is a brute face explanation, not an intentional one.

5 This example and the discussion in this paragraph were prompted by "The Time of a Killing," by Judith Jarvis Thompson, which appeared in *The Journal of Philosophy* 67 (1971): 115–32.

6 There are some cases where the purpose expression cannot be transformed so easily into the label for a complex act. If X does Y in order to obtain Z, it may be that obtaining Z is a completing act which successfully completes doing Y. In such a case, the complex act which contains doing Y and obtaining Z must be labeled in a more cumbersome fashion, as when we say that Smith competed successfully in the race.

7 I am ignoring the issue of determinism. Even if we suppose it is possible to formulate an intelligible version of determinism, which I doubt, the truth of that view would not make it incorrect to talk of possibilities, and we couldn't help doing it anyway. But it might make such talk less interesting or less important, for we could then say that all the possibilities that will be made actual are already "determined."

Chapter 8

1 The result is in principle due to Kleene [6]. A modified exposition can be found in [3, §13A] (for explanation of the references see the References. Note that in the theory of λ-conversion of a combinator is a λ-ob without constants or free variables.

2 The term was introduced in Kearns [5, p. 561] for combinators "which take account of their arguments." The definition in [3, §13C4, p. 270] is slightly different, but related. Hairsplitting exactness in regard to the use of this term is not necessary. The usage here will be specified more precisely in §2.

3 Thus Shabunin has a discriminator L (=Kearns R) such that for all obs X, Y, Z we may say $LX(YZ) = XYZ$. Then for $U = L\mathsf{K}$ we have $U(XY) = X$, which leads to contradiction by [3, Theorem 12A1. Note that the proof of that theorem requires the rule (μ).

4 I.e., combinators as understood prior to the introduction of discriminators. See below, §2.

5 Alternatively, by the Church-Rosser theorem for $\lambda\beta\delta$ conversion [2, §4D5].

6 For an introduction see [4]. A reader familiar with the first three chapters of that work can probably read the sections of [2 and 3] which are explicitly referred to.

7 In [3, §13C] this atom is called '⟦end⟧'.

8 Just how these numerals are defined is hardly relevant here. Throughout this paper they will be taken as abstract (in the sense of [3, §13A]); but a representation of them as pure combinators may be used instead.

9 This makes an n-tuple be of the form $\mathsf{D} + nX_1 \ldots X_n$ #. A somewhat similar idea occurs in [10], but the motivation is different.

10 These were ⟦lgh⟧ and ⟦concat⟧ in [3].

11 It is tempting to insert at this point the word 'essentially', thereby meaning to exclude such combinators as KK⟦Id⟧. But aside from the fact that this introduces a certain vagueness, it is preferable to allow that a discriminator may be equal to an ordinary combinator. For if we were to insist that discriminators form an equation-invariant class of combinators, then the undecidability theorem [3, §13B2] shows that it could not be a recursive class. (A similar remark applies to the arithmetical combinators of [3, §13A2].)

12 The standard reference is Markov [7]. A somewhat condensed description, sufficient for our present purposes, is in [1, §2E]. I shall follow the terminology of the latter. See also Mendelson [8].

13 Of course, we cannot use (17) if we use a terminal (16) to start the algorithm. If we assume that all preceding U_i contain at least one auxiliary letter, and that all V_i which are not stop orders (in which case they contain ω) do too, we can proceed as follows. Let β be an auxiliary letter not occuring in any command (extending L_1 if necessary); then insert just before (16) the commands $\xi \to \beta$ for all auxiliary letters ξ, $\beta a \to a\beta$ for all letters a in L_0, $\beta\beta \to \beta$, and $\beta \to \omega$.

14 Note that the absolute value $|n|$ is defined by primitive recursion from $|0| = 0$, $|n + 1| = 1$, i.e., $|n| = \mathsf{R}0(\mathsf{K}^2 1)n$.

15 Note that $\mathsf{Z}n$ and ⟦Id⟧an are σ-redexes in the sense of [2, §3D6].

16 Since Property (E) holds for $\lambda\beta\delta$-conversion by [2], Theorems 4C1 and 4C2, the cited theorem follows by a slight modification of the proof of [2], Theorem 4E1.

17 In fact, one can prove without too much difficulty a theorem saying that if X reduces normally to Y, M is a redex in X, and X' is the result of contracting M in X, then X' reduces normally to Y and the reduction is not longer than from X. Iterating this, one has the result in the text.

Chapter 11

1 The idea of intensional model structures originates with Saul Kripke. See Montague [7] and Scott [8] for general discussions of intensional model structures. For a methodological account, see Thomason [14].

2 A great many formal systems of conditional logic have been proposed,

provided with semantic interpretations, and proved complete in the literature. See Anderson and Belnap [1] for a survey of some of these. But few of these systems have actually been put forward as theories of the conditional used in any human language, and even fewer have been associated with any justification relating to the truth values of conditionals in a human language. In general, formal logicians have not given serious attention to this factor of justification. Moreover, problems arising with subjunctive or so-called "counterfactual" conditionals have not received much attention from formal system-builders until quite recently. For alternative contemporary theories see Åqvist [2], Lewis [5], and Manor [6].

3 See, for instance, Segerberg [9], Shukla [10], and Gabbay [4].

4 The definitions differ from those of [12] to ensure that $r(\Box A) = r(\Diamond A) = 1 + r(A)$. This change does not affect anything material.

5 These conditions are needed because, since partial interpretations are not defined everywhere we cannot presuppose, as in *D2.7*, that if $I_\alpha(A) \neq T$ then $I_\alpha(A) = F$.

6 The actual result of [12] applies to **CQ**, but the proof given there is easily simplified to the case of **CS**.

Chapter 12

1 Composition of this paper was supported by NSF grant GP21189.

2 The question of the *ordinal* of this consistency proof (which is evidently a transfinite induction) raises a curious point. Certainly we can define the system EBL of [2] in exactly Fitch's way and parallel his consistency proof verbatim in a system which is by ordinary proof-theoretic standards extraordinarily weak; for example, in the system IDB^ω, plus possibly some axioms of choice (see the very end of [8] for a description of this system). It is well known in the folklore that the ordinal of this system is the Bachmann ordinal $\phi_{\epsilon_\Omega}(1)$ (the result is presumably Howard's), therefore the ordinal of "extended basic logic" is even smaller than this. However, in order to show that, e.g., the system of [5] (without R6 and R7) is contained in that of [2] (cf. R34 of [5]), or to carry out the construction of elementary analysis in extended basic logic (cf. 29.9 of [5]), Fitch needs the highly nonconstructive principle that if for each b either ab or ~(ab) is provable, then either ab is provable for all b or ~(ab) is provable for some b. In other words, the fact that "extended basic logic" can be proved consistent by elementary methods and that if contains such-and-such a portion of analysis does not show that that portion of analysis can be proved consistent by elementary methods if *the fact that it does contain that much analysis* cannot be proved by such methods! A formalist, for example, would certainly accept Fitch's consistency proof of extended basic logic, but he would not accept for that reason the consistency of the system of [5] (without the implication rules), because he would not accept Fitch's proof that that system (with R34) is contained in EBL. The problem of consistency for the formalist is precisely the justification, by a consistency proof, of such things as what

we have just called a "highly nonconstructive principle," and the question of the ordinal of [5] (as distinct from EBL) remains, to my best knowledge, open.

3 There are, however, some other, later formalizations of considerable foundational importance, for example, the extensional systems of [6] and [7]. I very seriously recommend the study of these unjustly neglected papers. To this neglect I have myself, I am ashamed to say, contributed by some carelessness in my responsibility as a reviewer. In my review of [6] in *Mathematical Reviews* 23A (1962), pp. 279–80, I said that a certain contradiction (described by Fitch on p. 20 of [7] was derivable in the system of [6], which would imply that the consistency proof given therein contained an error. In fact, this criticism was sound as concerns the manuscript of Fitch with which I was working, but *not* as concerns the printed version, which is entirely correct. I can only hope that my unforgiveable gaffe has not done too much harm.

4 Two rules of the system of [5] besides the implication rules mention implication. We modify these by replacing the reference to implication by a reference to a subordinate proof. For example, the modified rule R44 will read "ϕc is a direct consequence of the following four items: $[R \text{ Pot } S]$, $[bSc]$, a subproof that is general with respect to y and has $[bRy]$ as its sole hypothesis and ϕy as its last item, and a subproof that is general with respect to x and y and has $[xRy]$ and ϕx as its two hypotheses and ϕy as its last item." A similar change is needed in R45 (in which, incidentally, $[x \cup z]$ is a misprint for $[x \cup \{z\}]$).

Chapter 13

The idea for this paper was conceived at the Summer Institute in the Theory of Knowledge at Amherst, Massachusetts, sponsored by the Council for Philosophical Studies in 1972. I am indebted to many of the participants for stimulation and criticism.

1 The possibility of thus reducing doxastic logic to *alethic* modal logic has already been explored by Smiley [34, pp. 113, 132] and Hilpinen [16]. Cf. also the reference to Levi [23, pp. 29f] in Hilpinen [15, p. 21]. For anticipations of Anderson's and Kanger's reductions, see Anderson [1, p. 170n. 38]. Kanger's definition had already been used by Fitch to get physical modalities out of alethic ones [12, §§11.19, 12.24] (as Anderson once pointed out in class). The sole difference is that for deontic logic, the additional axiom $\Diamond Q$ (Kanger) or $\Diamond{\sim}S$ (Anderson) is required, whereas for physical modalities, presumably L ("the physical laws," corresponding to Kanger's Q) must be assumed as a nonlogical (non-necessitable) axiom.

2 *Pace* Unger [35]. As Kripke points out, (~k6) would have the effect of identifying K with a *falsum* connective [19, p. 209].

3 In most such logics, (~k7) leads to L-omniscience, i.e., (semantically) the L-true is the L-truly known, or (metaphysically) "the real is the rational," as we might say.

4 For purposes of comparison only, I read $\Box$ or von Wright's V as *x knows that* (in contradistinction to *is verified*) and Hintikka's *defensible* as *satisfiable*. I would stress that the authors in question do not advocate these readings. von Wright would apparently be more at ease with a more serious system rejecting L-omniscience (necessitation) and (k4), or with an epistemized metalogic like Hintikka's; cf. [37, pp. 30f].

5 (b8) is redundant, following from (b1), (b3), and (b4).

6 In §§2 and 5 I omit brackets according to the conventions of Church as emended in my [5, p. 201]. Connectives are ranked, and dots to the *left* of two-place connectives take precedence over regular Churchly dots.

7 Since ($\sim$b2) and its dual $p \supset Cp$ yield $\vdash Bp \supset Cp$, Prior's proof that the latter is deductively equivalent (actually, interdeducible) with $M{\sim}\nabla$[29, p. 141] also establishes (b2). Cf. (b8) and n. 9 below, as well as Anderson [1, p. 173, 1-13] and Smiley [34, pp. 129f.].

8 ∇ corresponds to Anderson's $\mathscr{P}$[2] or B [1, p. 175], while ∇ corresponds to his S [1, p. 170] and Prior's $\mathscr{P}$[29, p. 140].

9 Besides $M{\sim}\nabla$ (n. 7 above), ($\sim$b8) has these interesting equivalents in normal logics: $C{\sim}\nabla$, Ct (t a tautology; pointed out by Smiley [34, p. 129]), $Bf \supset f$. If propositional quantifiers are added, it is furthermore equivalent to $\exists pCp$, $\sim\forall pBp$. If $\vdash BBp \supset Bp$ [or (pr) or (k4!) of §4] is added, another equivalent is $C\exists pBp$. If $\vdash Bp \supset BBp$, then $C\exists p(Bp \,\&\, p)$ is also equivalent to ($\sim$b8).

10 Updating Leibnizian theodicy, we might interpret it as the thesis that this is the best of all sets of possible worlds.

11 Cf. notes 21 and 24 below.

12 As Kripke inadvertently omitted the condition from his purported semantics for D2 [19, p. 220], what he had actually defined were C2 model structures. Cf. my [3].

13 The restriction to $\mathscr{N}$, which Routley adds, makes no difference to C2$'$, since in abnormal worlds (pr) is trivially true anyhow. The restriction would complicate our characterization of $\triangle$, however. Cf. notes 21 and 24.

14 The $'$ in C2$'$ is chosen as standing for the superlative of *prior*.

15 Cf. note 1 and (δ1).

16 Not valid in all normal deontic logics, nor in any normal epistemic logic. The quantificational counterpart holds also in the inclusive theory. Cf. (pr).

17 Cf. von Wright [37, p. 19 n.].

18 Cf. note 9.

19 Cf. note 20.

20 This follows by Kripke's Theorems 1-2 [19, pp. 206 f.] from the provability in C2 of $Bq \vee (Bp \supset Cp)$, an instance of Halldén's property.

21 On pain of some complication, we could have used Kripke's and Routley's original definitions of E2 and C2$'$ model structures, according to which $\mathscr{R}$ $\{\mathscr{S}\}$ need only be reflexive on $\{$right-reflexive from$\}$ $\mathscr{N}$ rather than its whole domain. Routley's restriction to $\mathscr{N}$ goes hand in hand with Kripke's. In that case, however, $I(\triangle) = (\mathscr{N} \cap \{h: h\mathscr{S}h\}) \cup \mathscr{M}$ for an arbitrarily chosen $\mathscr{M}$ disjoint from $\mathscr{N}$. By innocuously overspecifying $\mathscr{M}$ as $\{h: h\mathscr{S}h\} - \mathscr{N}$,

we arrived at our simpler interpretation. We pay the piper in resulting complication of Theorem 8; cf. note 24 below.

22 This does literally rule out a world in which x knows nothing but has beliefs. However, the intuitive content of such a situation could still be captured by saying that x knows nothing of a form other than $\triangle \supset p$.

23 In $C4'\triangle\dagger$ (= $C2'\triangle$ + S4 reduction + propositional quantification), ($\delta 2$) can be replaced by (t2) of §4 along with (pr). Some further interesting equivalences supplementing those of note 9 (which are also provable in $C2'\triangle$ or appropriate extensions): $\vdash \mathrm{Bt} \equiv \mathrm{B}\triangle \equiv \exists p\mathrm{B}p$, $\vdash \mathrm{Bf} \equiv \mathrm{B}\nabla \equiv \forall p\mathrm{B}p$, $\vdash \mathrm{Ct} \equiv \mathrm{C}\triangle \equiv \exists p\mathrm{C}p$, $\vdash \mathrm{Cf} \equiv \mathrm{C}\nabla \equiv \forall p\mathrm{C}p$.

24 These last two clauses are necessitated by the simplification of the characterization of $\triangle$ explained in note 21 above. The rest of the proof is considerably complicated thereby.

25 Argument-validity in the received sense of logical consequence. The "strength" of the completeness proof is not needed here.

Chapter 14

1 Double quotation marks are used in the same way that Hintikka uses them in [4]. They are placed around an expression to refer to any expression which results from it by replacing each syntactical variable in it by some expression to which that variable refers. Unless otherwise specified, the following conventions of Hintikka's are also used. '*a*' is a syntactical variable for names of persons, definite descriptions of persons, and personal pronouns. '*p*' and '*q*' are syntactical variables for independent clauses. '*B*' is an abbreviation for 'believes that' and '*K*' is an abbreviation for 'knows that.' '$\supset$', '&', and '$\sim$' are the standard propositional connectives.

2 Hintikka also carries out a parallel discussion of why it is odd for a person to say "*p*, but I do not know whether *p*," translating this as "$p \& {\sim} K_a p \& {\sim} K_a {\sim} p$." On occasion, he seems to have this same translation in mind for (4). It is clear that his key reasoning would be the same in both cases, however.

3 In [2], Frederic Fitch makes a similar point. He treats knowing as a truth class closed with respect to conjunction elimination so that, if *p* is true, but "$K_a p$" is not true, it is impossible that "$K_a(p \& {\sim} K_a p)$" is true.

4 The oddity of a person's saying any of the following statements could also be explained analogously: "*p*, but I do not think that *p*," "*p*, but I do not hear that *p*," "*p*, but I do not see that *p*," and "*p*, but I do not feel that *p*."

5 A more detailed analysis is provided in the appendix.

6 It seems interesting to consider what a might have done had he reflected on his argument as we have. Would he have decided simply to hope for the best or to try to believe for the span of each day that the event occurs on that day—an alternative that someone less argumentative might have chosen from the start.

Chapter 15

1 I am indebted to Risto Hilpinen for calling this point to my attention. His recent artible [6] independent of my own contains an interesting, but quite different, analysis of all these topics. My own analysis here is, however, analogous to his analysis of relativized alethic modalities [5].

2 Ruth Barcan Marcus has pointed out to me the likeness of such inferences—from (a) to (b)—to fallacies of composition and division and to fallacies involving adverbial modification. She also rightly suggested that collective obligations may properly be understood as binding upon ordered n-tuples.

3 Cf. Urmson [12]: "Simple utilitarianism, Kantianism, and intuitionism, then have no obvious theoretical niche for the saint and the hero . . . until so modified successfuly [to accommodate acts of supererogation] they must surely be treated as unacceptable . . ." (p. 67); and " . . . I do not think that the distinction of basic duty from other acts of moral worth, which I claim to detect in ordinary moral thought, is a sign of the inferiority of our everyday moral thinking to that of the general run of moral theorists" (p. 72); and also "To my mind this three-fold classification . . . is totally inadequate to the facts of morality; any moral theory that leaves room only for such a classification will in consequence also be inadequate" (p. 60). Urmson is correct.

4 Cf. Urmson [12]: "Traditional moral theories, I have suggested, fail to do this. It would be well beyond the scope of this paper, and probably beyond my capacity to produce here and now a full moral theory designed to accommodate . . . the facts of saintliness and heroism" (p. 67).

5 Cf. Urmson [12, pp. 62, 64–65, 73]. An example here would be a doctor who is truly a saint and, therefore, "volunteers to join the depleted medical forces in . . . a plague-ridden city" (p. 62) decimated by a germ bomb.

6 Current "impersonal" articulations of deontic modalities cannot even *express* this distinction—duties binding on all men versus duties binding on all men in a club, i.e., $O_U p$ versus $O_A p$—much less the distinctions *within* the membership of a club, i.e., $O_A p$ versus $O_B p$.

7 Further philosophical applications of my theory will, I hope, appear later. I will now only roughly indicate some directions for further work.

(a) Contrary-to-duty imperatives express what is obligatory if some obligations are not fulfilled. These are widely used in ordinary language and are important for ethical theory. As Chisholm (1963, p. 36) puts it, "most of us need a way of deciding, not only what we ought to do, but also we ought to do after we fail to do some of the things we ought to do." Contrary-to-duty imperatives escape adequate formalization in non-relativized systems of deontic logic . (For details cf. Føllesdal and Hilpinen [3, pp. 24–26.] If we relativize deontic modalities to classes of individual acts, the sort of theory presented here might perhaps adequately formalize "contrary-to-duty imperatives."

(b) The Neoplatonist and Athenian scholarch Proclus claimed that heroes belong to an ontological realm different from ordinary mortals.

Two questions arise. First, what precisely could such a presumably "bizarre" claim mean? Second, what conceivable support could it have? Both questions might perhaps be profitably pursued using the theory presented here. Indeed, I believe that my theory supports the claim of the celebrated Proclus.

Chapter 16

1 See, for example, Frederic B. Fitch, "On Kinds of Utterances," pp. 39–40 in Paul Kurtz, ed., *Language and Human Nature: A French-American Philosophers' Dialogue* (St. Louis: Warren H. Green, 1971). For fuller discussion of the motivation for the analysis presented in this paper and discussion of the relations between this analysis and the analyses developed by Belnap, Åqvist, Kubiński, and others, see David Harrah, "The Logic of Questions and Its Relevance to Instructional Science," *Instructional Science* 1 (1973): 447–67.

2 In various writings Belnap, Åqvist, Kubiński, and others have shown that much of erotetic logic can be developed without assuming the underlying language to have any special metalinguistic capability. For a discussion of motivation on this point, see David Harrah, "On Completeness in the Logic of Questions," *American Philosophical Quarterly* 6 (1969): 158–64. For further details on W*, normal interpretations, and communication applications, see idem, "Formal Message Theory," pp. 69–83 in Yehoshua Bar-Hillel, ed., *Pragmatics of Natural Languages* (Dordrecht: D. Reidel Publishing Co., 1971).

3 This construction is similar to the one proposed in Herbert G. Bohnert, "The Semiotic Status of Commands," *Philosophy of Science* 12 (1945): 302–15. Note that our analysis here is explicitly concerned only with the directive and the replies, not with the "inner logic" of F or the world state obtaining when F is satisfied. Thus this analysis is complementary to the one presented in Nicholas J. Moutafakis, "The Extensional Pragmatics of Commands," *Notre Dame Journal of Formal Logic* 12 (1971): 489–98. A similar point holds more generally for our analysis of other types of sentence, and our approach in general is thus complementary to the approach of R. M. Martin as in his *Toward a Systematic Pragmatics* (Amsterdam: North-Holland Publishing Co., 1959).

Bibliography of the Works of Frederic B. Fitch

1933. Note on Leo Abraham's "Transformations" of Strict Implication. *The Monist* 43:297–98.

1936. A System of Formal Logic without an Analogue to the Curry W-Operator. *Journal of Symbolic Logic* (*JSL*) 1:92–100 (dissertation in revised form).

1936. Physical Continuity. *Philosophy of Science* 3:486–93.

1937. Modal Functions in Two-Valued Logic. *JSL* 2:125–28.

1938. The Consistency of the Ramified *Principia. JSL* 3:140–49.

1938. Note on Hofstadter's "On Semantic Problems." *Journal of Philosophy* 35:360–61.

1939. Note on Modal Functions. *JSL* 4:115–16.

1939. The Hypothesis that Infinite Classes Are Similar. *JSL* 4:159–62.

1940. *Mathematico-Deductive Theory of Rote Learning* (with C. L. Hull, C. I. Hovland, R. T. Ross, M. Hall, and D. T. Perkins). New Haven: Yale University Press.

1940. Symbolic Logic and Behavior Theory: A Reply. *Psychological Bulletin* 37:817–19.

1941. Closure and Quine's *101. *JSL* 6:18–22.

1942. A Basic Logic. *JSL* 7:105–14.

1944. Representations of Calculi. *JSL* 9:57–62.

1944. A Minimum Calculus for Logic. *JSL* 9:89–94.

1946. Self-Reference in Philosophy. *Mind* 55:64–73.

1947. Remarks on the Theory of Types. *Mind* 56:184.

1948. On God and Immortality (English with Spanish abstract). *Philosophy and Phenomenological Research* 8:688–93.

1948. Reply to Professor Baylis' Criticisms. *Philosophy and Phenomenological Research* 8:698–99.

1948. Corrections to Two Papers in Modal Logic. *JSL* 13:38–39.

1948. An Extension of Basic Logic. *JSL* 13:95–106.

1948. Intuitionistic Modal Logic with Quantifiers. *Portugaliae Mathematica* 7:113–18.

1949. The Heine-Borel Theorem in Extended Basic Logic. *JSL* 14:9–15.

1949. On Natural Numbers, Integers and Rationals. *JSL* 14:81–84.

1949. The Problem of the Morning Star and the Evening Star. *Philosophy of Science* 16:137–41.

1949. Attribute and Class. In *Philosophic Thought in France and the United States*, ed. Marvin Farber, pp. 545–63. Buffalo: University of Buffalo Press.

1949. A Further Consistent Extension of Basic Logic. *JSL* 14:209–18.

1950. L'Attribut et la Classe. In *L'Activité Philosophique Contemporaine en France et aux Etats-Unis,* ed. Marvin Farber, pp. 220–42. Paris: Presses Universitaires de France. (French translation by F. Paliard of Attribute and Class [1949]).

1950. Actuality, Possibility and Being. *Review of Metaphysics* 3:367–84.

1950. Toward a Formalization of Hull's Behavior Theory (with Gladys Barry). *Philosophy of Science* 17: 260–65.

1950. A Demonstrably Consistent Mathematics, Part I. *JSL* 15:17–24.

1951. A Demonstrably Consistent Mathematics, Part II. *JSL* 16:121–24.

1952. *Symbolic Logic: An Introduction.* New York: Ronald Press.

1953. A Simplification of Basic Logic. *JSL* 18:317–25.

1953. Self-Referential Relations. In *Proceedings of the XIth International Congress of Philosophy* 14:121–27. Amsterdam: North-Holland Publishing Co.; Lowain: Editions E. Nauwelaerts.

1953. Justification in Science. In *Academic Freedom, Logic, and Religion,* ed. Morton White, pp. 99–107. Philadelphia: University of Pennsylvania Press.

1954. A Definition of Negation in Extended Basic Logic. *JSL* 19:29–36.

1955. The Reality of Propositions. *Review of Metaphysics* 9:3–13. (Also, in Japanese, in *Americana,* 2:59–67 [1956].)

1956. Recursive Functions in Basic Logic. *JSL* 21:337–46.

1957. A Definition of Existence in Terms of Abstraction and Disjunction. *JSL* 22:343–44.

1957. Combinatory Logic and Whitehead's Theory of Prehensions. *Philosophy of Science* 24:331–35.

1958. Representation of Sequential Circuits in Combinatory Logic. *Philosophy of Science* 25:263–79.

1958. An Extensional Variety of Extended Basic Logic. *JSL* 23:13–21.

1959. Quasi-Constructive Foundations for Mathematics. In *Constructivity in Mathematics,* ed. A. Heyting, pp. 26–36. Amsterdam: North-Holland Publishing Co.

1961. Sketch of a Philosophy. In *The Relevance of Whitehead,* ed. I. Leclerc, pp. 93–103. London: George Allen & Unwin Ltd.; New York: The Macmillan Company.

1963. The System CΔ of Combinatory Logic. *JSL* 28:87–97.

1963. A Logical Analysis of Some Value Concepts. *JSL* 28:135–42.

1963. Algebraic Simplification of Redundant Sequential Circuits. *Synthese* 15:155–66.

1963. The Perfection of Perfection. *The Monist* 47:466–71.

1964. A Geodelized Formulation of the Prediction Paradox. *American Philosophical Quarterly* 1:161–64.

1964. Universal Metalanguages for Philosophy. *Review of Metaphysics* 17:396–402.

1966. Natural Deduction Rules for Obligation. *American Philosophical Quarterly* 3:27–38.

1967. A Complete and Consistent Set Theory. *JSL* 32:93–103.

1967. A Theory of Logical Essences. *The Monist* 51:104–09.

1967. A Revision of Hohfeld's Theory of Legal Concepts. *Logique et Analyze* 39–40:269–76.
1968. A Note on Recursive Relations. *JSL* 33:107.
1969. A Theory of Computing Machines (with R. J. Orgass). *Studium Generale* 22:83–104.
1969. A Theory of Programming Languages (with R. J. Orgass). *Studium Generale* 22:113–36.
1969. Combinatory Logic and Negative Numbers. In *The Logical Way of Doing Things,* pp. 265–77. New Haven: Yale University Press.
1969. A Method for Avoiding the Curry Paradox. In *Essays in Honor of Carl G. Hempel,* pp. 255–65. Dordrecht-Holland: D. Reidel Publishing Co.
1969. *Dictionary of Symbols of Mathematical Logic* (ed. with R. Feys). Amsterdam: North-Holland Publishing Co.
1970. Correction to a Paper on Modal Set Theory. *JSL* 35:242.
1971. Propositions As the Only Realities. *American Philosophical Quarterly* 8:99–103.
1973. A Correlation between Modal Reduction Principles and Properties of Relations. *Journal of Philosophical Logic* 2:98–101.
1973. Natural Deduction Rules for English. *Philosophical Studies* 24:89–104.
1974. *Elements of Combinatory Logic.* New Haven: Yale University Press.
1974. Toward Proving the Consistency of *Principia Mathematica.* In *Bertrand Russell's Philosophy,* ed. George Nakhnikian, pp. 1–17. London: Duckworth.